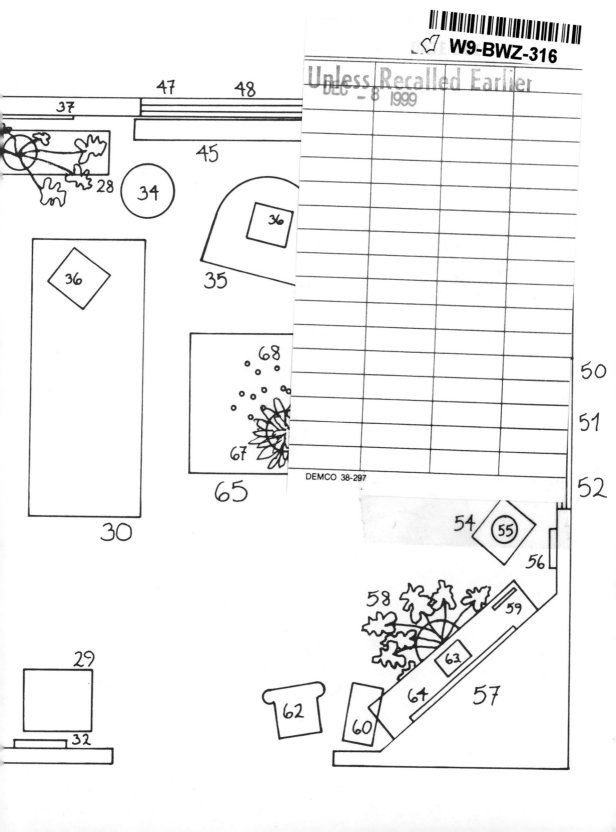

Home Rules

HOME RULES

Denis Wood and Robert J. Beck

with Ingrid Wood, Randall Wood, and Chandler Wood

THE JOHNS HOPKINS UNIVERSITY PRESS

Baltimore and London

All photographs by Denis Wood and Robert J. Beck, except as follows: title page photo and photos numbered 27, 46, 48, 49, 51, and 60 by Michael Zirkel; photo number 37 by Ron Rozzelle; and photos on pages 293 and 332 by Scott Sharpe, *News and Observer,* Raleigh, N.C., with permission.

Published in cooperation with the Center for American Places,
Harrisonburg, Virginia

© 1994 The Johns Hopkins University Press
Printed in the United States of America on acid-free paper

The Johns Hopkins University Press
2715 North Charles Street
Baltimore, Maryland 21218-4319
The Johns Hopkins Press Ltd., London

Library of Congress Cataloging-in-Publication Data

Wood, Denis.
    Home rules / Denis Wood and Robert J. Beck ; with Ingrid Wood, Randall Wood, and Chandler Wood.
        p.   cm.
    Includes index.
    ISBN 0-8018-4618-8 (alk. paper)
    1. Environmental psychology—Case studies.   2. Socialization—Case studies.   3. Living rooms—North Carolina—Raleigh—Psychological aspects—Case studies.   4. House furnishings—North Carolina—Raleigh—Psychological aspects—Case studies.   5. Family—North Carolina—Raleigh—Psychological aspects—Case studies.   6. Wood family   I. Beck, Robert J.   II. Title
    BF353.W66   1994
    155.9′45—dc20   93-19784

A catalog record for this book is available from the British Library.

In memoriam

ROLAND BARTHES

CONTENTS

ACKNOWLEDGMENTS

We would never have started this book had we not quite independently both found ourselves reading Roland Barthes's *S/Z*. That *sui generis* masterpiece authorized the very idea of our project, including its focus on a single room and its object-by-object and rule-by-rule slow reading. It also provided an example for the form of the book we ended up writing. While nothing Barthes wrote fails to find some echo in our work, the relentless and inclusive example of *The Fashion System* was a continual support. Other texts of Barthes particularly important in our work were *New Critical Essays* (especially "The Plates of the *Encyclopedia*"), *Roland Barthes* (especially the photographic essay that opens the book), and *The Responsibility of Forms*. To attribute to Barthes this or that aspect of our text would be to misjudge the nature of our dependence on his work and example. It is pervasive.

Almost as pervasive is the influence of Claude Lévi-Strauss. While our development of the binary oppositions in the room would seem to be due to his example, far more vital was the way he confirmed for us the interconnectedness of all things. Whenever we faltered, *The Raw and the Cooked, From Honey to Ashes*, and *The Origin of Table Manners* urged us on. When the road seemed unendurably long, *The Way of the Masks* and *The Jealous Potter* came along to insure us that the journey was worth it. And our phenomenological bent is probably as indebted to Lévi-Strauss's description of the sunset in *Tristes Tropiques* as to anything else.

A more important source for us for the system of binary oppositions in the room was actually Henry Glassie's brilliant and beautiful *Folk Housing in Middle Virginia*. To be sure, his sense of structuralism is deeply indebted to Lévi-Strauss, but while Lévi-Strauss heads you are often not sure where, Glassie sights a landmark and takes you straight to it. His incisive diagrams of Middle Virginia architectural logic continually clarified for us our aspirations. Glassie's work here directly informs our reading of every part of the "room." After we had completed the manuscript, our reading of Glassie's splendid *The Spirit of Folk Art* confirmed in us the adequacy of our reading of the room's art too.

Pierre Bourdieu's *Distinction: A Social Critique of the Judgment of Taste* enabled us to make explicit the relationship we felt between what Glassie taught us about the room and our own intuitive sense of class relations in contemporary American life, and this despite Bourdieu's exclusive focus on France. The form of the book—its subtle collage effect—helped confirm for us our decision to order our text as we have. Where Bourdieu was especially helpful in the social realm, R. D. Laing was essential in enabling us to think through the connections between it and the family. While Laing's *Politics of Experience* is deep in both our thought, *The Politics of the Family* was the text of his we turned to most frequently in writing *Home Rules. Reason and Violence*, which Laing co-authored with D. G. Cooper, was important as well. Irving Goffman's work was also critical. Again, *The Presentation of Self in Everyday Life* is anciently entangled in our thinking, but the notion of keying that he developed in *Frame Analysis* enabled us to understand Christmas and the other phase shifts the room undergoes; while the importance he places in *Forms of Talk* on studying social interaction in the whole environment was important for us, too. The sense we would like to convey here is one of our indebtedness to a whole world of thinking and writing: Kenneth Burke (*A Grammar of Motives*), Noam Chomsky, Guy Davenport (especially *The Jules Verne Steam Balloon*), T. S. Eliot, Paul Goodman, and Clifford Geertz (whose *Works and Lives* came out at a peculiarly propitious moment for us) are six others who come readily to mind. Conversations with David Caploe proved critical to our understanding of Marx, Lévi-Strauss, and Burke—that is, to the intellectual *structure* of our project.

Our notion of the *field* is derived from the physical sciences—from Faraday, Maxwell, Einstein. John Wheeler has been especially influential, and his *A Journey into Gravity and Spacetime* especially clarifying. What we understand by the notion has been enriched and extended by historians of science as diverse as Mary Hesse (*Forces and Fields*) and Donna Haraway (*Crystals, Fabrics, and Fields*) and by such scientists as Kurt Lewin (*Field Theory in Social Science*), C. H. Waddington (*Principles of Embryology*), and René Thom (*Structural Stability and Morphogenesis*), the cherodes and morphogenetic fields of the latter of whom are the forerunners of Rupert Sheldrake's morphic fields. We did not turn to this construct in the hope of lending to our work an air of the physical sciences—in fact, we reject the distinction between physical and social sciences (there is only one science no matter how many scales of analysis)—but because it emerged in our work as a powerful heuristic. Rupert Sheldrake's use of the construct in so similar a way was powerfully confirming, as was Pierre Bourdieu's (though to be certain his *espace hodologique* is closer to our sense of field than his *social space*).

Daniel Spoerri's *An Anecdoted Topography of Chance* constitutes a real precedent for our project. The appearance of his text in our acknowledgments is a sign that we reject not only the distinction between the physical and social sciences but that between the sciences and arts as well.

Jimmy Thiem inked the map of the neighborhood. Jill Beck, Roger Hart, Tom Koch, Arthur Krim, and Ron Rozzelle supported the project from its inception. Kelly Lentz, Garland Poole, and Frank Wellons politely answered questions about the rules of the house and otherwise enlivened things. Kelly was especially helpful. Anyone who has read this book knows it could never have been written without the cooperation of Nancy and Jasper Wood, but the contributions of Marie Krawchek are less evident. Denis's paternal aunt, Marie, was always at the other end of a phone line or letter to answer questions and provide leads. Christopher and Peter Wood and John Hansen also answered questions and volunteered suggestions. The importance of the roles played by Ingrid, Randall, and Chandler Wood are attested to by the presence of their names on the title page, but the aid provided by Lide Anderson, Carrie Knowles, Mary Meletiou, and Susan Parry in helping Denis and Ingrid deal with Ingrid's breast cancer during the writing of this text would go unremarked unless acknowledged here.

One of the points our analysis of the room stresses is the coherence and wholeness of the world in the unendingness of its filiations. That we were enabled to write this book ultimately must be attributed to the unfolding of the universe at large. This book is the way the universe expresses itself—through our lives—in the here and now.

INTRODUCTION

The Living Room

What is a room? And if this seems too silly or trivial a question, what is a room that when we enter it we all but invariably know where to go and what to do? When we wrote this book, we had not read Rupert Sheldrake's *The Presence of the Past: Morphic Resonance and the Habits of Nature,* but the answer we have worked out to our question about a room is not unrelated to the answer he has come up with for the questions he asks: What makes a rabbit rabbit-shaped? How do newts regenerate limbs? Why are molecules shaped the way they are? Why do societies arrange themselves in certain predictable patterns?

Sheldrake's answer is that memory—in the form of habit—is inherent in nature. He suggests that a beech seedling grows into the tree that it does because it inherits its nature from previous beeches, not simply through its chemical genes, but from past behaviors to which it "tunes in." If this is what happens—to us as well as to beeches—then there is no need for such "habits" to be stored in our nervous systems. "The past may in some sense become present to us directly. Our memories may not be stored inside our brains, as we usually assume they must be."

We are not sure about all of this, or even about what some of it means, but we conceive of a room in somewhat the same way Sheldrake conceives of a beech, as the instantiation of a kind of collective memory (we say this frequently: *the room is a memory*). What we mean by this is that the room is a mnemonic, certainly: it stores *in the arrangement of its parts* how, for instance, we will sit with each other so that we do not have to figure it out every time anew. But, more fundamentally, we mean that, insofar as these arrangements emerge from the pasts of those living the room, the room is a *memory* of those pasts. In the living room we study in this book, Denis and Ingrid Wood's children encounter the memory of the living rooms in which Denis and Ingrid grew up, which in turn were memories of the living rooms in which their parents grew up, and so on. In Sheldrake's example, "as a swallow grows up, it flies, feeds, preens, migrates, mates, and nests just as swallows habitually do. It inherits the instincts of its species through invisible influences, acting at a distance, that make the behavior of past swallows in some sense present within it."

Among the influences bringing the past into the present in humans are the dos and the don'ts parents use to give form to the behavior of their children. With respect to rules, we conceive of such rules . . . *as rooms in the form of rules.* That is, whatever the room *is*—memory, mathesis, economy, organ, culture (and we imagine it to be all of these)—it exists not only in the material substances out of which it is composed but also necessarily and inevitably in the form of the behavioral regularities we transmit to our children in rules.

It is true that we also transmit such behavioral regularities through example, as well as through the behaviors "implicit" in the arrangement and structure of the walls and furniture (through something like the affordances of J. J. Gibson or the morphic resonance of Sheldrake). But the rules are methodologically privileged. Whereas the behaviors are only implicit in the physical structure, and difficult to observe and record directly, the rules are both explicit and easy to collect, vivid as they are in the minds of children and parents of a certain age. We have collected the rules for the living room of our study from Randall and Chandler Wood—then 11 and 9, respectively—as well as from their parents. We did this by having Robert Beck ask the boys to tell him the rules for the couch, the walls, the leather chair, the windows, the glass table, the door. (Subsequently he collected rules from Denis and Ingrid, too.) In this fashion, 228 different rules were collected for the seventy "objects"—cabinets, house plants, fireplace—into which the room was split up. We imagine these rules articulating a "field"—again, not wholly unrelated to Sheldrake's "morphic field"—one capable of determining behavior in a manner analogous to the way a magnetic field determines the path of a subatomic particle hunting its way through a particle accelerator.

If for kids this field has to be made explicit in the rules, adults respond intuitively to it implicit in the room. The field is actually one of values and meanings, *values and meanings* embodied in the ceiling and the floor, the house plants and the door. That is, the things of the room embody the values and meanings that made, selected, arranged, and preserved them, values and meanings that the presence of kids makes explicit as rules. For us, then, a room has three forms, each of which necessarily implies the others: that of the values and meanings (the room expressed in value and meaning space); that of the rules (the room expressed in rule space); and that of their material embodiment (the room expressed in physical space). Each of these forms can be viewed through any number of "filters" (those of culture, memory, economics, education, connoisseurship, and so on), and because none is privileged we attempt to look through as many as we can.

Since for us the values and meanings, the rules and their physical embodiment, are merely alternative forms of one thing (the room), we do not see the room as a setting for a life but as one of many ways in which that life expresses itself. As the life unfolds, the room it expresses unfolds too. The life the room expresses now—while we are writing this introduction—is not the same life it expressed six or seven years ago when we first started studying it. Then Kelly Lentz, who in the pages of this book will be found occasionally hanging around the house, did not live with us as he does now, did not sleep in what used to be Denis's office, and did not do most of his studying in the living room. But this transformation, while unforeseen, was not unprepared for. That the living room is now bright with Kelly's learning is an expression of precisely the values and meanings we *originally* found in studying the room (that is, that the room became Kelly's classroom was a potential expression of its values and meanings). Just as Sheldrake's beech seedlings are prepared to grow into trees, the room was prepared to unfold into a classroom, because of the values and meanings (because of the past behaviors) into which both are "tuned." Since for us a room is but an aspect of a life, just as we speak of "living a life" so we speak of "living a room." The "living room" we are studying is living thus not merely in distinction to kitchen or bedroom, but because it really is.

The Plan of This Book

The book has two interwoven parts, a tour of the room and a series of comments—divagations, really—on it.

The tour proceeds object by object (screen door, door, doorframe, window in the door, and so on). Each object is presented in three ways, corresponding to the different forms in which we see the room. First, the object is displayed in its physical form, by means of both a photograph and a verbal description. Here we attempt to capture some sense of the extent to which even the most taken-for-granted aspect of the room (a wall) embodies both the material sophistication of a technologically advanced culture (gypsum plaster, titanium dioxide in vinyl acetate) and a powerful phenomenological presence (in the sunlight it . . . glows).

Then the object is exhibited in rule space. Here we simply list the rules collected for the object at hand. For example: RULE 74: *"Don't put your hands on the walls"* (Chandler, tAPP), RULE 75: *"Don't mark up the walls"* (Denis, ktAPP), and so on, noting who volunteered the rule (here Chandler and Denis) and in which of the great concerns of the rules (appearance [APP], control [CON], protection [PRO]) it is most involved and to which domain it most pertains (t for the thing itself, k for the kids).

Finally the object is explored in value and meaning space, entry to which is gained by using the rules as keys (here: "What are the walls that one should not put one's hands on them?"). Throughout each discussion the object and its rules are related to other objects (denoted by arabic numerals, usually in parentheses, and keyed to the map of the room on the endpapers) and their rules (denoted by arabic numerals following RULE in small caps) to indicate how the rules and objects conspire to generate the value and meaning system of the room as a whole; how, that is, they form a net or felt of meanings and values capable of shaping the behavior that gives rise to the room (which embodies the values and meanings [which shapes the behavior {and so on}]). The relationship here is akin to that articulated by John Archibald Wheeler when he observes that, "spacetime grips mass, telling it how to move," but that "mass grips spacetime, telling it how to curve."

We imagine the effect of this object-by-object and rule-by-rule reading of the room to be that of entering and scanning the room . . . *in slow motion*. The divagations (in roman numerals) comment . . . *on these processes*, that is, on both the process of scanning the room and living it (the first is naturally embedded in the second). These comments—these divagations—arise in the text as we imagine they might arise in any attempt to come to grips with the significance of the room (with its values and meanings). Because they arise in situ, they are probably best read as they appear in the text, but they can be skipped over and returned to later or ignored altogether (it will, to be sure, be a different book). In fact, the book can be read in any order. It was written to be read from beginning to end (from the screen door to the light switches), but with only a little more effort it can be read from the inside out (starting, say, with the wicker rocker) or at random. Or all the physical descriptions can be read in sequence, followed by the rules, the values and meanings, and the commentaries. The plan of the room, table of contents, rulelist, and index are provided to facilitate access to the book and to the room it attempts—in a sort of fourth form—to embody.

Home Rules

I · THE FIELD

What is home for a child but a field of rules? From the moment he rouses into consciousness each morning, it is a consciousness of what he must and must not do. If during the night his pillows have fallen to the floor, he must pick these up, for *pillows do not belong on the floor, they belong on the bed.* If he thinks of turning on the radio, he must keep it low, for *we do not play the radio loud before everyone is up.* If he needs to urinate, he must go to the bathroom, for *we put our wee-wee in the toilet.* If he is old enough to stand, he must lift the seat, for otherwise he might splash and spatter, and *we do not do that in this house.* When he has finished he must flush the toilet—for *we always flush the toilet when we're done*—and lower the seat—*because that's how we do things around here.* He is not to sing gloriously in welcome of the day, nor dance a fandango back to his bed, nor wake his brother by eagerly whispering in his startled ear, "Quick, Watson! The game is afoot!" because *we don't wake people up until they're ready*—unless they're kids and they've got to go to school. Then the rule is, *you've got to get up in plenty of time for school!*

So many rules! No matter how you count them, the number is enormous. Is it one rule that the spoon must go to the right of the knife, and another that the knife must go to the right of the plate? Or is the way we set the table one rule altogether? Either way, the number of rules about no more than the way we eat, where we eat, when we eat, what we eat, and who eats with us is alarmingly large. Around these, like electrons about the nucleus, swarm still others, rules about how we come together to eat (for instance, *with clean hands*) and rules about how we dissolve the meal *(may I please be excused?)* and still others about washing the dishes and putting them away and who cleans up the dining room and when and how thoroughly, though it is difficult to say which rules swarm around which others—hierarchies are hard to see through the haze of rules, and those that at first blush seem superordinate often turn out to be no more than vague *(you know better than that).*

Hundreds of rules? If the meaning of rule is taken narrowly *(those spoons go in the drawer to the right of the stove),* there are more likely thousands. Yet without them the spoons might end up anywhere, *would* end up anywhere, out in the sandbox when it's time to eat, or down at the bottom of the creek, though why would anyone care, dinner would be . . . dinner *wouldn't* be, there wouldn't be any dinner, no sitting down together, no shared breaking of bread, no shared gulping of milk *(if you're going to gulp your milk like that, you can just go out to the kitchen).* Without the rules the home is not a home, it is a house, it is a sculpture of wood and nails, of plumbing and wiring, of wallpaper and carpet.

II · REACH

The question that the teacher asked the student, "You wouldn't do that at home, would you?" takes as granted the universality of the rules, which, we pretend, either make sense or are natural, or both, and so, like sunlight, are always and everywhere the same. Rules are justified as rational—it is demonstrated that failure to comply leads to catastrophe *(if everybody talked at the same time, no one would be able to understand them)*—or as no more than simple codifications of human behavior, the way everyone has always acted *(a man never wears a hat inside a building)*. That the very necessity of saying the rule denies the foundations on which these justifications are erected is ignored or denied: but how else to expect obedience without coercion? Or else there *is* recourse to coercion *(if you're not going to do what I say, you can just go up to your room)*, the issue of justification is never broached, and rules are emitted as edicts ex cathedra, often precisely because they make no sense *(I know you don't like what they gave you, but you have to thank them for it anyway)* or because the natural way is condemned *(I know it's difficult, but you must control yourself)*. Nevertheless, whether rational, natural, or arbitrary, the expectation always is that they will provide guides for the conduct of actions everywhere *(we expect you to be as well behaved at Greg's house as you are at home)*.

But at the same time it is widely acknowledged that when in Rome one does as the Romans do—or at least that when having dinner at the Schaffners' one does as the Schaffners do *(if they insist on saying grace, the least you can do is wait until they're done before you start eating)*. But it's not just the Schaffners: other homes are legion, there are many Romes, and any number of places, times, and conditions under which the rules are expected to be adapted, bent, or canceled *(yes, but you're supposed to use your head)*. Nor is it only that everyone agrees that the behavior appropriate to the beach might be inappropriate at a ball. It is that there are rules observed at school which are not observed on the street, rules observed in the playground which are not observed at home, and rules observed in the living room which are not observed in the bathroom. The pretense to universality crumbles under the demands of every specific site (culture is concrete: it is not manifested in general, but necessarily in situ), and although people *say* the rules, the rules are embodied in specific actions and things. To enter a room is to find oneself immediately amid objects whose character and arrangement admit only of certain possibilities, it is always to enter a unique system of rules. The rules educed from this room often may be exogenous (most will be), but inevitably a number will prove to be sui generis and the ensemble will be a singular property of the time and the place. To enumerate the rules completely may suffice to define any room, just as Maxwell's equations fully characterize the electromagnetic field.

III · TO STUDY THE RULES . . .

Although the opposite doesn't follow—the enumeration of things and actions pertaining to a room will probably not exhaust the set of all rules (just as Maxwell's equations are incapable of describing quantum and gravitational fields), such an enumeration seems a good place for an environmental psychologist and a psycho-geographer to commence their study of rules, with those of a room. There is a substance to a room, however elusive it may prove to be: a concreteness, a solidity, a finitude. The room may connect to other rooms, but at some point you know you have left it, and so have left the field of rules of which the room is no more than a manifestation. There is a beginning and an end to such rules, to the rules of a place. They can literally be enumerated (there are only hundreds of them): they can be known. And they are known. Those whose room it is know the rules, they have imported them from other rooms, or they have invented them for this place alone. No rules will have been introduced by our scrutiny of the room, and in fact the whole system is naturally occurring (which is not to say that it is not of culture)—not merely the room and the rules, but the articulation of the room's rules *(you know you're not supposed to eat in the living room)*. It is as though we had been handed a text to read: whatever the character of our reading, the specification of the text would not enter into it: there it would be, delimited by its incipit and its explicit. Our text, however, is a room and its rules as spoken by those whose room it is. We intend a sort of environmental ethology, where the attention to the room removes the rules by at least one degree from those enouncing them, permitting us to see the rules in action, as it were, not as aspects of personality.

IV · . . . IS TO STUDY THE ROOM

But what is the room?

To say, "To enter a room is to find oneself immediately amid objects whose character and arrangement admit of certain possibilities," is not to lie, but only to unfold sequentially (that is, in time) an experience of simultaneity (that is, of space);

for one does not first enter a room, second, perceive objects in it, and third, attach significance to them (language makes us do that), but one enters all at once this room/objects/significance-thing, this culture (the room is a culture), this mathesis (the room is a way of coming to know the world), this organ (it is a functional structure in the organization of the house), this curriculum, this mnemonic, this field of forces through which one moves as a proton through the magnets of a particle accelerator:

"Please come in," and it may be your hostess speaking, but it may be the way a rug lies on the floor that whispers this invitation.

"Do sit down," and it may be your host who says this, but it may be a certain capaciousness to a sofa that draws you in and gathers its cushions around you.

"Look at this."

"Be careful of that."

"Watch your feet."

"Make yourself at home."

They are *heard* in time—these voices are heard in time (their text unfolds over many visits)—but the volume of air and the quality of light do not speak independently of the sofa and the table, or of the way you are drawn to sit down and smile at your hostess. The walls and floor do not speak before, or after, the rest of the room. They are not apart, not other, from the rest of the room. The room without objects and the meanings they shed is another room, it is not this one, it is not the one we entered. In our human living, the petty kingdoms ruled quite independently by architect and decorator and sociologist have no independence: it is not the painted plaster alone that sings to us, but whether something hangs upon it. And if it is a mirror, the song is other than if it is a painting or a print or a calendar whose nudes cradle in their creamy arms replacement parts for pumps and fans.

Which is not to deny the walls, not to deny the volume of air, not to deny the doors or the windows or that quality of sunlight which at any rate at night ceases pouring through them. We have no wish to deny the room in the experience; but to reduce the experience to the room is to leave life for architecture, as it is to leave life for the social sciences when the experience is boiled down to its meaning, or to move into a furniture store when all that is regarded are the furnishings. Even to refer to them as furnishings is to miss the point, for few rooms are furnished; they are not stages set by a designer on which some actors will recite their lines, they are not settings: they are the resultant—in the sense of a sum of vectors—of a living.

V · JOINTS

Let us establish a convention. It is artificial, but so is the language it is intended to circumvent. When we refer to "a part of the inside of a building, shelter, or dwelling unit usually set off by partitions"—that is, to the room in its exclusively architectural aspect—we will put the word inside quotation marks: "room." But when we wish to refer to that resultant of a living inclusive of this "room," the things in it and their significance, we shall leave off the quotation marks and just write: room. Our desire is to articulate this room—here, in these pages—to give it joints.

We live the room.

We spend time, do things, argue, eat, make love in the "room," but we live the room. The room is not apart from our living and is no more a consequence of our living than our living is a consequence of the room. Like the outer integument of a pearly chambered nautilus, the room is a shell secreted in the ongoingness of the organism's perseverance. Though it may be common to think of the shell as empty, this is only because we typically encounter shells after the death of their creator-inhabitants, after weeks and months of scouring by sand and water. The shell comes from the sea, it is a *sea*shell; we want to imagine it has nothing to do with cephalopods, with gastropods, that it has to do with the water exclusively: we do not see the beach as a charnel house. It is as though we were to encounter rooms exclusively in the roofless labyrinths of the ruined palaces at Knossos and Mycenae, as though we were to take as rooms none but the bat-infested hollows at Chichén Itzá and Uxmal. Yet even these are mnemonics, even here we can hear the echo of the rules.

These rules are a form of the room, just as the room is a form of the rules. They *are* the room, expressed in rule-space (where there are also rules expressive of things other than rooms, things like courtship, dinner, and city streets). This is not a game of words: to know the rules of a room is in some sense to know the room (incompletely, but then complete knowledge of the room comes only from living it), just as to *see* the room at some moment in time (perhaps from behind the velvet ropes of the guided tour) is in some sense (incomplete again) to know the rules. "I wonder," one muses, "what they did *there*," but what they did *elsewhere* is perfectly certain (the rule is obvious) and there is little question about the mass of rules in most of the rooms we are daily in. The rules are—we would say—*self-evident,* and we follow them without reflection that this is what we're doing. Among adults the rules are explicitly stated only with respect to extremes *(I think I should warn you that's extremely fragile),* snares *(be careful walking on the rugs, they're very slippery),* and illusions *(you can sit in that chair if you wish, but let me warn you that the bottom is likely to fall out);* or as an opportunity to tout a value that might otherwise escape notice *(the only reason I caution you is because it's four thousand years old, and*—deprecating laugh—*a little difficult to replace;* where what is understood is, *the only reason I caution you is to provide myself an opportunity for mentioning its age and thereby also displaying the attractive modesty which forbids that I mention what I paid for it,* where what is understood is, *I have class, taste, and resources).* Otherwise the rules are not so much taken for granted as unexpressed, potential, latent in the values, latent in the significances we find in the objects and the "room," with respect to which the rules are but an expression (as is the character and arrangements of objects themselves, as is the room).

An expression of values: *the room is an expression of values,* the room that is also a shell secreted in the ongoingness of the organism's perseverance. What can this mean but that values too are a shell secreted in the ongoingness of the organism's perseverance?

Nothing, it can mean nothing else: the values do comprise a shell, a shell which, like the room, is not something bought and moved into (bought at the Value Store, democratic values down the aisle, autocratic values on the third floor, Christian values—on sale—in the basement); a shell which is not a setting acquired once and for all within which mental life takes place *(where did you get your values?)* but is exactly like the room, a resultant of a living. It is, however, mental, known to the organism summing the values but to no one else except insofar as they are bodied forth in action. To say this is to deny the values nothing except corporality, is to imagine them much as we imagine a magnetic field which, whether or not made visible in the filings of iron, is nonetheless there: similarly the room makes the values visible, embodies them in *its* corporality, *is* the values in their corporal form.

This is not to say that there exists between the room and the values it incarnates an isomorphism, or that the room may not constitute itself an attempted deception, fraud, or illusion. The expression of values in a room is neither certain nor univocal. It is constrained by the available resources (the room is an economy), or the values are self-contradictory and the room must represent one position at the expense of the other (the room is an interpretation), or the room variously embodies the values (the room is a performance), or it embodies yesterday's values (the room is an archeological site) or the values of one's parents (the room is an echo, a memory, a shrine). Moreover, the room may be constrained to express the values of more than a single person, and one expression may be garbled by that of another, or unnaturally reinforced, or completely canceled. Or a novel expression may be observed, the resultant purely of the simultaneous expression of unrelated values: a room may embody the values of both husband and wife, for instance, each imperfectly, and the resultant expression may be of a certain sociability or sense of compromise. Or a rented room may express the values of both owner and occupant, and thus lend substance to no more than an empty and conceivably unintended sense of subordination (as when the bare walls of the occupant substantiate nothing but the owner's proscription of putting holes in them).

And yet, despite the infinity and subtlety of these filters, the understanding is that all adults can see through them, that all adults can apprehend the values behind them, that all adults will respect the room accordingly, that is behave appropriately, as though they shared the values concretized by the room. This means that explicit verbalization of these values is unnecessary—they are obvious, *they are written in the room*, and consequently such explication inevitably takes on the air of an act of patronage, as though the individual receiving the explicantia were someone who might not see, might not apprehend, that is, as though he were not quite an adult, as though he were—perhaps—a child.

VI · STRIPPING THE ROOM

Because it is not presumed that children will be capable of reading these values, the presence of children in a room precipitates a veritable orgy of rule enunciations: *don't touch, be careful, that's fragile, stay away from that table, don't sit in that chair, that clock's not something to play with, NO RUNNING IN THE LIVING ROOM!* It is like lowering the cathode into an old-fashioned battery: immediately a stream of electrons is stripped from the anode. So too the introduction of kids into a room strips rules from every space or object the kids approach, rules dissociate themselves from walls and floor *(I don't want to see any dirty hand marks on these walls; don't just leave your stuff all over the floor)*, from glass surfaces and wooden ones *(please keep your finger-prints off the windows; you know putting your wet glass down there is going to leave a ring, why do you do that?)*, from individual objects and the room as a whole *(you are not to touch those speakers; when people are talking in the living room, you don't just barge in and interrupt)*. No part or aspect of the room escapes being implicated in this reaction whereby values are transformed into rules. To put it slightly differently, kids *excite* adults to express the values manifested in the room as—in the form of—prescriptive rules. This enables the adults to *maintain* their values (by protecting the room from the barbarians, from the kids) at the same time that it enables them to *reproduce* them (that is, to instill the values in the barbarians, to inject them into the kids through the hypodermic of the rules). A room without kids, or other barbarians—that is, without someone deemed incompetent to respect the values implicit in it (of which it is always a bodying forth)—is a room whose rules are latent. Any member of a similar culture, that is, anyone who has lived a similar room, may be presumed able to imagine what the rules would be were the room to be invaded by nieces and nephews, by the kids from down the hall, by the sons and daughters of a colleague, by a neighbor's . . . dog.

But until such invasion, the rules remain precisely that, imagined: potential utterances, future imperatives, conceivable curses, but, inevitably, latencies. Because the presence of kids excites these into action—stripping them from the values they prescriptively represent—the presence of kids exposes the values in their verbal form to the scrutiny of any observing intelligence, no longer through the presumptive transparency of the objects and their arrangements within the "room," no longer through "a reading," but as though they were naked: the presence of kids forces the rules to disrobe *(Aw, mom, why do we have to?).*

In other words, to flush the values from their cover of furniture and rugs, sparkling windows and unsmudged woodwork, all that is necessary is to strip away the rules. And all that is necessary to strip away the rules is the presence of a kid . . . or two. Two kids—the sons of one of us, of Denis (and Ingrid)—helped us with this work: Randall, then age eleven, and Chandler, then age nine. Over the years their presence in the living room had stripped away somewhere between three and four hundred rules. Denis and Ingrid knew these rules: they had *actualized* them, turned them into speech for, as they would say, "the benefit of the kids" (but actually for the benefit of the room and, often enough, for the "room"). However that may be, these rules were directed to the kids—are directed to the kids; it was their behavior these rules were expected to shape, and Denis and Ingrid often violated the rules, or construed themselves as exceptions to them, by virtue of age, or experience, or special circumstance. While it is probably true that adults continue to "grow up" through the process of teaching these rules to their kids—because the presence of the kids in general embarrasses them into obeying them (or else into sophistry, or into shouting [coercion], or into junking the rules)—nevertheless, the room is *a field of rules* essentially for the kids. And so, while the other of us (Bob) did inventory the room and its rules by interviewing Denis (Ingrid reviewed this work, commenting on it as she saw fit), he elicited the majority of the rules comprising the corpus for our reading from Randall and Chandler.

VII · PARSING THE ROOM

To inventory the room: what can this mean?

Only this: that the room will be broken up, disassembled, shattered. Of necessity, this must be an arbitrary operation. If the room *is* a shell secreted in the ongoingness of the organism's perseverance (in this case, the family Wood's), then its fundamental characteristic *as a room* lies precisely in its wholeness. Although women's magazines of a certain class make no bones about hacking up a room—a Regency table on a Moroccan rug, a Parsons table beside a leather sofa—they never do more than list furnishings (they express fashion): what (barely) looks like a room—what wants to look like a room (but there is no life in it)—is often not even a "room," and frequently the subject of their enumeration is literally a *set* organized exclusively for the provision of a photograph which their prose will then *establish*. These cleavages that seem so evident at a certain distance (plant, table, couch) are less apparent closer up, while others that previously went unnoticed suddenly seem unavoidable: it is a matter of the level of attention, but to say this does not reduce the terror of the hanging question: where to cut?

That plant over there, for instance, all at once reveals itself as a collection of plants, each plant dissolves into: a plant, potting material, soil, pot, saucer, coaster. That table, a simple pure whole if ever one existed (it is Mies van der Rohe's Barcelona table), immediately shatters into legs and glass top, four white plastic studs, and a small transparent disc (which often gets lost) used to compensate for a floor anything but flat. The couch explodes into a throw pillow, three back cushions, three seat cushions, and the couch proper, each with its own slipcase—that is, into sixteen parts, each in turn susceptible of further division (for example, the main body of the slipcover has 118 individual snaps). Nor is this division at all academic: in the life of the room the couch gets dirty, the slipcovers have to be removed, they are washed, hung out to dry, and replaced. Before the floor can be waxed—twice a year—the table has to be moved. To do this, the glass top is lifted up—it is not attached to the legs, it just sits on them (this is its elegance)—and put someplace (often on the couch, itself already moved), while the legs are put somewhere else (anywhere). Inevitably the little transparent disc is misplaced. The plants are moved all the time, they get taken in and taken out, are moved to locations of more sun and less sun, get repotted, die (and though this often seems to be the only thing they do by themselves, they also live—this is their charm). In any case, today's ensemble is not the one that occupied its location the day before (the ensemble in the corner by the speaker now—as we write—is not the one that was there when we inventoried the room).

At the same time, plant, table, and couch are caught up in the coffee table ensemble, to sit on the couch is *always* to confront this table, and beyond it the plants *(some* plants in any event): couch, table, plants comprise a tissue in this room that is an organ in the life of the house, a whole no less whole than the couch is a whole (the proposition is ambiguous), than *its* cushions are wholes, than *their* slipcovers are wholes, than *their* closures are wholes, than *their* snaps are wholes, than *their* stud sides are wholes (they have been manufactured, you can pick them up, look at them, you can sew them to the cloth), than *their* sewing holes are . . .

Perhaps here, with the holes in the stud side of the couch-cushion slipcover snaps, we have lost it (there is nothing left: the coffee table ensemble has been analyzed into holes), lost the room, this room anyhow, the living room, for with the pushing of the needle through the sewing holes of the stud side of the snap we have entered another room, the room in which Ingrid made the slipcover, perhaps, or one in an upholsterer's shop, but we are no longer here, we are somewhere else.

Where to set the distinction? Where to draw the line? Where to cut? Snaps? Cushions? Couch? Over by the coffee table? Living room? It is not as though you could get to one without hacking through another: any cut you make cuts through everything; to get to the heart you have to cut through skin, slice through muscle, saw through bone; and then the heart is only the literal heart, it is not the one that will answer all the questions, it is not the one you were looking for, it is just the machine that pumps the blood, dead outside of its position in the structure, dead and meaningless. So it is with the room: one cuts not for understanding (not to find the essence), but for convenience (to get on with the living): *don't slide down the banister!* and though it might puzzle Randall and Chandler to say what a banister is, they understand exactly what they're not supposed to do. And this is the way *we* cut up the room, without thinking too much about it, completely arbitrarily (completely "naturally," we would like to say, but we know better), as though we were walking through it, saying what we saw:

"Screen door" (but not stile, rail, screen, handle, latch, lock).

"The door."

"The doorframe."

"The window in the door."

"The windows in the sidelights" (but we overlooked the fanlight).

"The bells" (the string of camel bells hanging from the doorknob).

"The latch and lock."

"The floor."

These things (we shall refer to them as things) sometimes imply a system (in referring to the large record cabinets, for example, we embrace in that phrase the twelve fourteen-inch Palaset cubes, the six Palaset bases, the Palaset connectors, and the 960 records, sleeves, and associated jackets), other times parts of systems (thus we distinguished among the newel post, the banister, the stair treads and stair risers): it was a question of creating someplace from which to start, a place from which to launch ourselves into the rules, and except insofar as the things had had rules stripped from them by Randall and Chandler, their nomination was a matter of essential indifference.

VIII · THING AND RULE BY THING AND RULE

Things. Things and rules stripped from them by Randall and Chandler. These are, of course, the room, turned through a certain angle, but it is also how we shall proceed, thing and rule by thing and rule. This unfolding will follow a kind of order (of necessity it will follow *some* order), but this order will never be presumed. Though we shall start with the screen door, the door, the doorframe, the window in the door, the windows in the lights, and the bells, and so work our way into the room, we shall never imagine that the room starts with the screen door, the door, the doorframe, never imagine that the room might not be entered from the dining room, or from the second floor via the stairs, or—on waking up from a nap—from the couch, from within the very "room" itself. Immersed in a book, one looks up and . . . enters the room.

Herein lies the advantage of our procedure, for halting as it is (pausing as it does to admire the sparks struck from each thing, the small fires kindled by the mingled rules), it gives as a result no conclusion, no place to stop, no . . . larger structure (it gives no canonical Room, no general Rule, no archetypal Thing). Instead of closing the door on the room—which *is* the things, which *is* the rules (which is the *values* energizing the rules)—it throws open the windows as well, allowing the admitted breeze to touch first this and then that, to keep the room stirred up, to keep it from solidifying, to prevent it from precipitating *the* crystalline room, *the* collection of objects, *the* sociology of rules. There are only *these* rules and *these* objects and *this* "room."

There is only this . . . plurality, this plurality compounded of momentary injunctions, arbitrary parsings, edges which are contingent at best. Yet echoing in this plural—in this room—are voices, of which the rules are no more than the most explicit articulation. It is these voices to which we must attend—to which we *do* attend—for it is these voices that give us, not the room (there is no room, only its living), but the dimensions of the space in which it is lived: there is the Voice of Comfort (soft couch, old wicker rocker), the Voice of Convenience (rugs up in the summer), the Voice of High Culture (the records, the paintings), the Voice of a Certain Easy Formality (an implicit asceticism, its violations), the Voice of a Well-Ordered Life (the neatness, the cleanliness). Through these, behind them, underneath them—it is not easy to tell where they are coming from—still further voices can be made out—

that of the bourgeoisie certainly, but also others, the voice of a dead modernity; more than a whisper of Anglo-American puritanism; perhaps, if you listen attentively, the murmur of a smothered anger . . . unless we are trying too hard and mistake but a buzzing in our ears.

Each voice speaks more or less loudly, more or less insistently, and if in this instant it is the Voice of Comfort which dominates *(all the comforts of home),* then at another it is that of High Culture *(when taste tells the tale),* and in any event as much depends on those who listen as on those who speak. To denominate these voices is not to delimit them or to hear in their commingling any particular harmony, but to point out a few of the voices whose presence is demonstrably audible through the codes which organize the rules. Mies van der Rohe's table, after all, is but a table, forever capable of slipping back into this identity, this transparency of a simple function: "It's just a table," one can say. "An orange crate would serve as well. You put stuff on it and it holds it up." But no sooner has the table all but been absorbed in this simple utility, than the rules force it to reappear in all its many other roles:

"Don't leave *Tintins* on the table."

"Don't play with anything that will scratch it up."

"Try to keep your fingers off the table."

"Don't lift the glass part off."

"Don't put your feet on the table."

"Don't leave your stuff on the table."

Why not? the kid asks and whatever the answer it is clear that the voices heard dimly through these rules are anything but those of Comfort and Convenience. Civilized folk can play the civil game of subtle deceptions in which none are taken in but all pretend to be, but the imperatives educed by kids cut through even honest ambiguity, slice through to the parent rock from which all the rest is made. It is this property that has determined our dependence on the rules, not merely our study of them: in their utterance, preserved against our forgetting in the perfect memories of kids, the mask is dropped: the world otherwise so transparent *(it is only a coffee table)* and natural *(what else could you put there?)* is suddenly opaque *(but what* is *a coffee table?)* and cultural *(how about a fire pit?).* Linking the rules through their codes to the voices which speak through them points to a network of consistencies that is no more than another way of saying *culture,* no more than another way of recognizing and reaching the boundaries of a freedom which is essentially illusory and in any event entirely contingent. Or is it real, but unused? It makes a difference, this distinction, but only at the level of the parent rock. One's culture may be constructed, not inherited, but given the culture, the rooms we live are granted too, these rooms which for children first and last are fields of rules, fields of rules established by the things whose nature (as we say, but really whose culture) is enounced by the voices whose timbre is most clearly caught . . . in the rules.

And what does it matter where one starts as long as the start is made?

The Screen Door

To approach the house on a very nice spring day is, for most people, to approach the screen door. It stands there in its white anodized aluminum thinness and lightness, the only thing between the inside of the house and the world without. Two fiberglass screens—occupying the upper three-quarters of the space between the rails (the lower quarter is filled with a kickplate)—are spring-loaded into an aluminum frame hung from hinges integral to an aluminum doorframe. This, in turn, is screwed to the wooden frame (3) from which the door proper (2) is hung. The black metal push-button latch—equipped with a thumb lock—catches a strike plate quite independently screwed to the wooden jamb. An automatic door closer was once fastened to the mullion strip at the top of the kickplate, but all that remain are the four holes in the mullion strip, the four holes in the jamb, and a cloud of memories. A spring and chain door stop—often unlatched to permit the door to be fully opened—connects the upper rail of the screen door to the jamb of the main frame. The absence of a hold-up spring means that the chain dangles there in plain view.

The fiberglass screening itself is no longer tight in its frames but sags and billows where the hundred hands of little kids have pressed against it. The splines frequently work out of their channels in the frames and have to be pushed back in with the business end of a key. It is a working door. In the winter, storm windows replace the screen and the membrane that had been permeable to most everything passes little more than light and heat and the muted sounds of traffic.

rules
1. "Don't push on the screen" (Chandler, tPRO).
2. "Don't push things through the holes in the screen" (Ingrid, tPRO).
3. "Don't slam it" (Chandler, ktAPP).
4. "Close it every time you go through" (Randall, tCON).
5. "Don't kick it" (Chandler, tPRO).
6. "Don't open it to strangers" (Randall, kCON).
7. "Don't open it to strangers or certain others" (Denis, kCON).
8. "Don't talk through the screen to guests or friends" (Denis, kAPP).
9. "Don't lock the screen against family members" (Denis, kCON).

1

So many rules?

But then, the thing is present to us in *so many ways*. It is, in the first instance, inevitably, a thing—that is, a material body. As such, it falls under the sway of an entire cacophony of degrading processes that reach conclusion only in the thing's destruction: the thing is cracked, ripped, dropped, smashed, abraded, worn, spotted, scratched, smudged, torn, warped, bent, broken . . . in a word, ruined. Unless it be preserved, that is, unless it be repaired, mended, picked up, put back together again, sanded, replaced, washed, polished, cleaned, straightened, set, fixed—all of which could be avoided (parents unceasingly promise their children) if only the causative acts could be prevented, that is, if the thing could be protected from the barbarians, from the kids. And vice versa, for each act of destruction slashes two ways: it is not only the window that is broken (with its attendant anxieties of having to get a new piece of glass—*and when am I supposed to do that?*—and of having to pick up some caulking compound, and then of having actually to fix the window), but the kid's skin which is (sickeningly) cut (with its attendant nightmares of having to staunch the flow of blood—without getting any on the couch—while trying simultaneously to calm the hysterical and guilt-ridden child *and* find the hydrogen peroxide, gauze, tape and *the damned scissors which nobody ever puts back!!*); it is not only the objet d'art which is smashed, but the kid who falls on his face among the shards; it is not only, to get to our case, the screen that is gradually ruined by the tiddling fingers, but the mosquitoes and flies that subsequently make it inside to buzz around, annoy, and ultimately bite. So in the first place, the thing needs to be protected from the kids, but no more than they need to be protected from it.

When a rule has this function in the regulation of the living known as the room, we indicate it thus: PRO, noting that it is more or less oriented toward protecting the *thing* with a prefixal *t* (tPRO), more or less oriented toward protecting the *kid* with a prefixal *k* (kPRO).

The second way the thing is present to us is in precisely its utilitarian aspect: *it keeps the flies out.* Here it is not that it *is* that matters, but what it *does:* and what it does is keep out, reflect, insulate, support, let in, permit, terminate, hold, cushion, enhance, moderate, modulate—in a word, control—the processes that are a living incarnate. In that these processes are oriented either toward the child more directly—toward *his* body, toward *his* behavior—or toward the thing more directly—toward *its* maintenance, toward *its* use—control, like protection, has to be viewed under two lights. *Close it every time you go through* is a rule oriented more directly toward the door: it says what one should do to the door *(close it),* it implies a host of things that may or may not be kept out of the house, it isolates in a very broad way the role of the door: to keep things out, to act as a filter of the stream of stuff flowing toward the house. *Don't open it to strangers,* on the other hand, is directed through the door to the child, is motivated by an image of what could happen to the child were this one particular thing (a stranger) to gain entry to the house, though actually it cannot be intended to achieve even this end (it cannot be intended to save the little pigs from the wolf), for in the actual case it can never have been imagined that Randall or Chandler's refusal to open the screen door could impede the entry of a determined stranger, given that the door proper is wide open, that the fiberglass is sagging in its frame, and that the screen door itself is unlocked. Thus the rule is not only *not* directed toward the door, it is scarcely directed toward the child except insofar as it exploits the door as a prop in his *education,* helping teach him to discriminate among strangers, certain others, guests, friends, and family. Because it cannot *work,* the rule is forced to play; it is a form of playing about control. It is: *playing control,* part of a highly elaborate game, *playing house*—only this is not the game kids play among themselves, but the much more serious one parents play with a child in the attempt to reproduce their culture (to instill it in the barbarians, in the kids).

Rules concerned with control we indicate with a CON, prefixing the *t* or *k* as needed to suggest—nothing more is possible—that it leans more toward the thing or more toward the kid.

But how to code *don't slam it?* On its face it begs to be taken as a rule for the protection of the door. Slamming cannot be good for the door: therefore we protect the door by prohibiting slamming. On the other hand its patent subject is a matter of control, what one *does* with the door: one opens a door (to let in) and closes it (to keep out). There is no room for slamming in this scheme, and to preserve the scheme we prohibit slamming: it is a way of defining the door (it is something we do not slam).

In fact, these are detours whose purpose is to naturalize the cultural, for the prohibition of slamming is devoted neither to the door as a thing nor to its role. It is concerned, rather, with the way one *is* toward the thing. It is a matter of relations, of attitude, of orientation, of style, of appearances—and so it falls under another code, the appearance code (APP). Such rules are the least likely to frankly advertise their allegiance. This is expressly because they are the most difficult to justify on grounds of nature (that is, the most difficult to justify to kids without invoking parental will). On this count they are those rules most likely to lead to the heart of the culture, for which very reason they are those most likely to camouflage themselves as rules of protection or rules of control. In fact, screen doors are to be not slammed because either it is irritating *(if they slam that door one more time . . .)* or else it bespeaks kids entirely lacking a sense of decorum, the awareness of which can be no less irritating *(those jerks!).* In both cases what is offended is not the door or its function (after all, the whole point of a screen door with a spring stop is to close itself, and in the absence of the automatic door closer this is certain to be with a WHAM!), but some (often implicit) sense of *the way things ought to be.* That is, what is offended is culture in its purest manifestation, in particular that Voice of a Well-Ordered Life that is so surely the hallmark of a certain class of professional. Or, to be more explicit, there is in the slamming of the door a kind of abandon, almost a wantonness, that Denis, at any rate, cannot accept in his children (Ingrid stresses the noise—that is, the disregard for others): nonetheless, this is a question of appearance. And the question of appearance is what most profoundly motivates the ban on letting the door slam, so much so that, although the sound per se of a distant door slamming is not only accepted but positively appreciated, insofar as such slamming is a sign of a certain lack of self-control, the signified pollutes the signifier, and the sound (the door's) and the act (the kid's) are both subjects of the rule. We indicate this by prefixing to an APP both a *k* and a *t*.

It so happens that the nine rules elicited for the screen door refer to the three codes under which all the remaining rules can be grouped. That this can be so in spite of an undeniable marginality invariably associated with screen doors (there is nothing, for instance, of Architecture about them) reflects the wider significance of the screen door in the home—to say nothing of the house—where this home may be understood as having the same relation to house that room bears to "room." That is, though we may play games, have dinner, do homework, sleep, and do housework in the house, we *live the home,* in the living of which we also live rooms, usually, though not necessarily, in "rooms." When the door proper is open, the screen door regulates access to all four of these: (1) to the "room" (in order to get into it from outside), (2) to the room (to collaborate in its living), (3) to the house (necessarily, for the "room" is located here, but also because this "room" is a passageway to other "rooms"), and (4) to the home, here not a collection of "rooms" or even rooms, but a living in its own right. Thus the distinction between inside and outside made by the screen door for the house is paralleled by that between insid*er* and outsid*er* made by the screen door for the home. More narrowly, the house may be characterized, among other ways, as an association of domestic fauna—one largely commensal—whose distinctiveness is guaranteed by the relative impermeability of its edge: roof, walls, floor, foundation. In this edge—let us refer to it as a membrane—the screen door acts as an important gating device: it permits us and all the truly microfauna through, while stopping the birds and the bees, the moths, mosquitoes, and flies, the stray cats and the neighbors' dogs. *It keeps the flies out*—hence the simple rule, *close it every time you go through.* At the same time, it defines *us* for the home, specifies, that is, the macrofauna for whom the home is a living. This is less simple, and so, where one rule sufficed for the house, four are needed for the home. *Don't lock the screen door against family members*—emitted to stop the kids from locking each other out (and adventitiously their parents)—has nonetheless the force of defining the family: it is those you do not lock out of the house. Denis's father once locked the kitchen door against his son, when, as we say, Denis was still living at home (but what can this phrase mean?). Denis got in through a window, but the sense of violation can be reanimated by any locked door: the urgency with which he advances this rule is born of this memory. The memory awakes other memories: he was returning from a tryst; the overhead light was on; Nancy and Jasper had been fighting; Denis and Jasper shouted at each other. If the room is a mnemonic, so is the screen door.

Don't talk through the door to guests or friends needs to be completed to make sense: *you invite them in!* Family you cannot lock out, friends and guests you cannot keep standing out on the porch. On the other hand, strangers are those you have to keep out (*don't open it to strangers*)—that is, *strangers are those you do not let in.* The *certain others* (of RULE 7) comprise a class that comes and goes, swells, shrinks, and vanishes. It includes neighbors (often children) whose presence inside violates someone's sense of the house (often Randall's or Chandler's). At the time we elicited the rules, Frank and Billy were personae non gratae. At the time of this writing—six months later— Billy is present in the "room" (he is characteristically annoying) and Frank is by way of being the kids' friend. The *certain others* form a twilight class: when they get invited in, they'll have become friends, or, they'll have become friends when they get invited in: it happens like that, suddenly, at the door.

We feel at a door—even a screen door with sagging screening—the full force of the rules of the room, of the home. It is all here. The door is a plentitude. It is certainly a mathesis. No less obviously it is a functional tissue of an organ of the house (if the room were a stomach, the screen door would be the esophageal sphincter). It is a mnemonic. It reflects an economy (the screen door is cheap). It is part of a curriculum. The field of forces it projects protects it. It establishes and maintains a system of appearance. It regulates certain functions of the house and in so doing generates a taxonomy of persons, a taxonomy based neither on sex nor size nor color nor age, a taxonomy established . . . by the door. Few eat in the house, and fewer still at the kitchen table. But among those who do, fewer still *just walk in.* Whatever their kinship, these who do are family—that is, collaborators in the living that is the home.

And all the others are . . . something else.

The Door

The door is not only from without: it is also an aspect of the room. When closed it is part of the wall; when open it is a sculptural presence. It is a large, heavy object, the

largest, heaviest regularly moved object in the house. Despite the glass occupying the better part of its upper half, the door is solid and reassuring. It closes with a comfortable *thunk*. It is a mobile piece of wall.

It has been constructed from many pieces, the glass, its frame (4), and the three molded panels caught up in a framework of stiles and rails. Mortised into the side of the butt stile are the leaves of three brass hinges. Each is attached to the stile with four screws. The other leaf of each hinge is mortised into the jamb, through which its screws have been sunk into the jack stud hidden behind the casing (3). Bored through the lock stile is a hole for the cylinder of the lock, perpendicular to which a hole is bored for the plunger and latch bolt. A latch-bolt plate is mortised into the end of the stile and secured with two screws. Over the hole bored for the cylinder lie the interior and exterior roses from which the doorknobs protrude (7). The strike plate for the latch bolt is fastened to the other jamb. We would say mortised but this is not the first strike plate to be emplaced and the jamb is a shambles of gouged wood and old hardware. Still, it works. The satisfying thunk of the closing door is compounded of two sounds. The higher pitched is that of the latch scraping across the tongue of the strike plate hole. The lower is contributed by the door falling against its stop.

There is a certain roundness to this sound provided by the insulation. V-strip weather stripping has been nailed to the jambs of the door just within the stop, while around the stop itself a metal strip edged with plastic tubing has been fastened. A bottom sweep has been attached to the bottom of the door and the original threshold has long since been replaced by a threshold weather strip. There is a tightness to the opening and closing of the door, a rightness that complements its weight and solidity. The door is opened, it is closed: *thunk*.

rules
10. "Don't slam the door" (Denis, ktAPP, tPRO).
11. "Don't play with the door" (Denis, kPRO, ktAPP, tPRO, tCON).
12. "Don't use the door as a 'gate' in a game: play inside or outside" (Ingrid, kPRO, tCON).
13. "Don't get your fingers caught in the door" (Denis, kPRO).
14. "Don't hang on the door" (Denis, tPRO).
15. "Don't kick it" (Chandler, tPRO).
16. "Don't open it too wide because it scratches the floor" (Chandler, tPRO).
17. "Don't let just anybody in" (Denis, kCON).
18. "Don't open the door to just anybody" (Denis, kCON).
19. "Don't go to the door and when you see somebody's there, just go away. Go up and open the door and see who it is" (Denis, kAPP).

20. "Close the door against the heat" (Ingrid, tCON).
21. "Close the door if the furnace is on" (Ingrid, tCON).
22. "Close the door behind you" (Denis, tCON).
23. "Open the door for fresh air" (Ingrid, tCON).

In fact, the screen door is a fake.

That this diaphanous poseur is able to project the rules of the room is entirely due to the authority of the door that stands behind it, ready at the yank of an arm to make substantial what in the screen is no more than gossamer. The screen door is not even skin: it is more like clothing, changed with the seasons; and in fact, here, it is literally clothing, assumed with the aluminum siding in which the house was clad in the early 1950s. The door proper transcends skin: it is all but bone: it is cartilage.

It is the door, then, that is actually the plenitude, the curriculum, the translucent elastic tissue that generates a taxonomy of persons. The screen door? It is opened *just to get to the door.* Denis and Ingrid tell a story. They were at dinner with the kids in the kitchen. They heard the door open (they heard the bells, they heard the whinny of the hinges). It had to be Martie: who else would . . . just walk in? "Martie!" they called, "Come have dinner!" Instead of Martie coming toward the kitchen, they heard feet begin the ascent of the stairs to the second floor. Denis rushed to the living room to find a stranger on his way back down. "Yes? Can I help you?" "I'm looking for Herman. He said, just come in, up the stairs, first on the right—but that's a bathroom!" "Yes," said Denis, "probably—but he was talking about the rooming house across the street."

And explaining himself, Denis showed him the door.

What else could he have showed him? There is nothing to *show* in a screen door (one would have to have said, "And explaining himself, he showed the stranger *out*"): air is its essence. But the essence of door is earth, it is bone. A door is bone that stretches: it is wall that moves. Whereas the screen door could keep the flies out, the door proper can keep out the very air, and thus *we close the door against the heat* and *close the door if the furnace is on.* Here it is acknowledged that as the house is an association of domestic fauna, it is also a bubble of air the quality of which is zealously monitored: it is *a little cool (don't you think so?)* or it is *quite warm* and *who turned up the thermostat?* or *sniff, sniff, there is a strange smell around here somewhere* or *I'm opening up: it's too stuffy in here.* As the single largest aperture in the membrane assuring the integrity of this bubble—at twelve square feet the door is nearly as large as the effective openings of the three operable living room windows taken together—it is the single most powerful element in its control; and so while, if already closed, *we always close the door behind us,* we also *open the door for fresh air.* The issue is control of the *house* in a simple and direct kind of way (the door is a valve) just as the issue is control of the *home* when the rule is *don't let just anybody in* and *don't open the door to just anybody* (where the door materializes a taxonomy of persons).

But as person is not independent of speech, so here control of the home is not independent of control of the house, for to make himself known a stranger needs to make his voice heard, but where air is blocked, so is voice (the door is a gag). Thus, a problem of appearances, here that of a guarded hostility: the doorbell rings, Chandler heads toward it but failing to recognize who is there retreats—back up the stairs, out to the kitchen, remarking, "I don't know who it is." The subsequent approach of Denis or Ingrid, trailed by Chandler—who is, of course, curious (he was only obeying RULE 18: *Don't open the door to just anybody*)—can only be that of the security guard summoned to superintend the after-hours breaching of the walls of a downtown office building as someone working late attempts to let himself out. This obsessive scenario is predicated on the perceived threat of a hostile environment, and its inherent defensiveness projects onto the stranger at the door—one of Denis's students, some-one intrigued by Ingrid's plants—the offensive character of a thief, of a child molester, of a proselytizer. All of which contradicts the cordial and generous reception Denis and Ingrid want to imagine they extend to guests and strangers (that they are hospita-ble, that theirs is a hospitable home, even that their living room is welcoming). Hence: *you don't go to the door and when you see somebody's there, just go away. You open the door and see who it is.* Yet, *you don't open the door to just anybody.*

Except for the empty accommodation achieved through abandonment of the latter rule as a consequence of the increasing maturation of the kids, these rules, rooted in different worlds (a world of utopian amicability, a world of resigned hostility), can-not be reconciled, or can be reconciled only in the ancient ambivalence surround-ing *strangers;* but in any case their simultaneous enunciation represents decidedly more than the transition from a phase in the living of the home during which the door is answered exclusively by the adults to one in which the children have become adults: it represents precisely the precarious suspension of a predisposition toward hospitality—in which the doors are opened to strangers (etymology: hospitality can be manifested *only* toward strangers)—in a world shrill with the conviction that strangers are never more than potential enemies (etymology: strangers *are* enemies). In the attempt to inculcate friendliness but not foolishness, one doesn't *open* the door to strangers (RULE 6) but does open the door to *see* who it is (RULE 19), where for the kids the greater ambiguity induced by the nonconformance of the rules is lost in the lesser ambiguity investing the word "see," where, as they point out, given the window in the door (4), "We can always *see* who it is." In fact, it is only when they can *see* that they do not know the person at the door that they are urged to open it to *see* who it is, where what is elided in the synesthesia is an interest in the stranger's purpose, in his or her *intentions*. What the conflict demonstrates is that if Denis and Ingrid are not Philemon and Baucis, they want to be, not in fear of losing their house in a flood of divine wrath, but out of a desire to see in their home a temple (etymology: consecrated place). Here the door becomes a marker, not between sacred and profane, but between the sacred and the cursed.

Yet at the same time—or, more precisely, at any *other* time—the door is an invitation to play. This is inevitable: there is a string of bells (6) hanging from the doorknob (7). The doorknob turns. The door swings open. It can be slammed shut (it will be slammed shut to materialize the screen door). It can be hung on. It can be swung on. Things can be crushed between the jamb and the butt stile (the door is a giant lever, it is a nutcracker). It can be played behind. This is especially true when the door stands open, when having broken away from the wall to invade the room— sweeping out six square feet of floor—it hangs with all its parts temptingly exposed, begging to be fiddled with, creating between it and the wall of the room a small wedge, a cave, in the corner. One might crouch in this corner to surprise another coming through the doorway, who might in reaction attempt to flatten the first by shoving the door back against the wall. Butt against wainscoting, desperate to escape Flat Stanley's fate, the first pushes back with all the strength he has coiled, the second steps out of the way, the door kicks forward, the first stumbles and his fingers slip . . . between the butt stile and the jamb. Or the second is shoved off balance and his fingers get caught . . . between the jamb and the lock. It is this hysterical scenario that Denis rehearses, recalling with a physical thrill the pain of the car door closing on his fingers—or were they, as his brother Peter insists, *Peter's fingers,* and it was Denis closing the door? In the calculation of sibling debt it matters, but here it only matters that playing at the door propels Denis into a sort of mad-dog fury of rule enunciation, *NO PLAYING WITH THE DOOR* and its obsessive variants, *don't use the door as a "gate" in a game— play inside or outside, don't get your fingers caught in the door, don't hang on the door, don't kick it*—which even Chandler notes are all "basically playing with the door" and which Randall observes "apply to every door in this house." It is the memory of a car door on West 25th Street in Cleveland—and perhaps here no more than the memory of a *premonition*—or the suppression of the memory of a bedroom door in the apartment on Spruce Court (and the responsibility for Peter's pain), which through its infection of *all* doors, erupts through *this* door into *this* rash of rules.

IX · THE VOICE OF COMFORT

What for the child is a field of rules is for the adult a nest of comforts. From the moment he rouses into consciousness each morning, it is a consciousness of what does and does not satisfy. The pillow is comfortable beneath the head or it is not. The room has a fine fresh air or it is stuffy. The floor feels good to the feet or it is unpleasantly cold. The coffee is wonderful or something is wrong with it. The shower is great . . . or not. This is the Voice of Comfort, graciously acknowledging each little satisfaction or (less graciously) its absence, an aperitive voice, bespeaking, finally, the endless contrivances of bourgeois society toward whose luxuriance of satisfactions everything in the home conspires, including the door, or, perhaps . . . the door above all.

Strictly speaking (from this perspective especially), there is no *door,* there are only the doors: the distinction between them is empty (or it is spoken in a different voice), the rules apply equally to both, or make sense for either only given the other, or migrate from door to door with the seasons. Door and screen door work together, through their control of the doorway cooperating in a vast collusion of walls and windows and doors and fans that is organized to maintain the integrity of the living called the home—and not incidentally a certain level of comfort—by controlling the permeability of its edge.

Each of the parties to this operation is ridiculously simple. The door proper has but a single operating condition (this is its seriousness), and if only because the house is not level and when open the door swings *all the way open,* two states, open and closed: it is never cracked, it is never ajar. The screen door has two states too, but it also has two operating conditions (this is its frivolity): its opening is glazed or screened and it is open or closed. The prevailing combination of states and conditions is a function of sets of simultaneously applicable rules and habits with dominion over three distinct scales of operation. At the nearest, a decision is made to open or close the door(s) in response to the pressure of a sharp spike on some perceptual index: Martie is at the door (and *you let her in*), Ingrid has driven up in a cab filled with groceries (and *you keep the doors open until you've brought them in*), smoke is pouring from the kitchen (and *you open the doors to let it out*), a jackhammer is raising a racket in the street (and *you close the door to keep it out*). At a less immediate—we would prefer to call it *local*—scale, the door stands open or closed as a mark of the family's receptibility. Here the encysting system expands to include porch, living room, and other house lights, the telephone, and information about habits of receptibility circulated among friends and acquaintances: the house is *closed up for the night* (and the door is closed) or Ingrid *goes to lie down for a while* (and she closes the door) or Denis wishes *not to be disturbed* (and he closes the door) or, after the kids are in bed, the adults decide *to go for a walk around the block* (and they take the phone off the hook, cut off the living room lights, and close the door). At the furthest, or global scale, the door is open or closed and the screens or storm windows put in or taken out according to a handful of rules that regulate the quality of air inside the house: *we close the door if the furnace is on* (that is, in the winter) and *we close the door against the heat* (that is, during the day in the summer), but *we open the door for fresh air* (that is, summer mornings and summer nights, and spring and autumn days).

The door is thus subject to a threefold sovereignty. Immediately the door is under the sway of the taxonomy of values it materializes (stranger/acquaintance, food/trash, fresh air/smoke, noise/quiet), but these are not solely of the moment, and the door serves also to maintain the climate of the house (the quality of its air) and the climate of the home (is it "family time" or can Frank come in and play?). The door *is* a valve but it is also a sign of receptivity, albeit one whose signified is able only intermittently to inhabit it as the signifier undergoes its seasonal metamorphosis: in the winter, when the door is closed to retain the heat diffused by the radiators (9, 45, 46), *that* it is closed means little (there is no difference in which to root the meaning). Then, when the days are shorter and the dark more prevalent, it is the state of the living room lights—the lamp by the couch (34) and the one by the plants (42)—that constitutes the sign. But with the onset of spring the door and the operable windows (47, 50) will be more and more often open during the day, though it will be some time (a measure of the uncertainty associated with seasonal change) before the glass is taken from the storm door. Only then will the door proper come into its own as a valve on the local scale and hence as a sign of the family's willingness to receive visitors. But with the onset of summer, the door proper will be increasingly closed again during the day, as will, slightly later, the operable windows, and slightly later still, their storm windows. At summer's height the door will be opened early in the morning but soon closed, along with the windows and their storm windows, none of which will be opened again until evening (though even then the door itself will be closed as soon as the whole house fan is activated). Again, *that* the door is closed when it is all but always closed means little at the local scale where the sign of receptiveness operates (at the global scale it is widely accepted as signaling the presence of air conditioning). With the onset of fall, door and windows will be more and more often closed at night and opened later and later in the morning. One cold day—often Halloween—the screens in the screen door will be replaced by the storm windows. The door itself will be less and less likely found standing open. Slightly later, the storm windows in the windows will be more or less permanently closed for the season. It is a kind of endless thermal farce, one window being raised as another is lowered, this fan off, that screen up, these storm windows down, those windows down, the furnace on, the furnace off, windows up, storm windows up, screens down, fans on, doors open, windows closed, storm windows disappearing into the basement, screens popping up from the basement—it's a *dance,* a door and window dance (with fans: it's a fan dance), it's a glass, wooden, aluminum, and fiberglass round dance, a dance around the seasons, a dance around the days, a dance done to the song sung by the Voice of Comfort.

But what is this Comfort achieved with so much opening and closing of windows and doors? Certainly it is no bed of roses. Certainly it is not the life of Riley. It is a labor. More precisely, it is the *form* of a labor, the form of a living: Comfort is the shape of an exertion. This shape is anything but innocent. Not only is Comfort bought with great effort (and Comfort is always a *great* effort), but it is bought at great *expense.* No more than the enumeration of the glass, wood, aluminum, and fiberglass consumed by the doors is required to sketch an entire economy that Comfort justifies; but when it is acknowledged that the fans run on electricity generated by the burning of coal and the fission of uranium, and that the furnace burns oil, immediately the climate of the house is implicated in the climate of the globe, from the changes induced directly by the increased concentration of CO_2 in the atmosphere to those secondary, tertiary, and quaternary consequents not exclusive of the presence of American warships in the Persian Gulf. The system of contrivances that is the Comfort of this home reaches from the assemblages of stiles, rails, hinges, screws, latches, and weather stripping that is the door; or the concatenation of belts, switches, motors, and whirling blades that are the fans; out, through the extended structures of appropriation, into every pit and plain of the planet's economy.

Not only is Comfort the consequent of a massive contrivance, but its ends are not simple, they are not straightforward, they are not direct. For instance, it is the first requirement of True Comfort to appear, precisely, uncontrived. The image is invariably that of the mother and father and son and daughter lounging in their comfortably heated living room while outside the snow falls; it is never that of the elaborately concealed system of ducts and pipes, heating machinery and fuel storage structures; never that of the oil man coming to fill up the tank for the winter; never that of the still vaster (but no less elaborately concealed) system of pipelines, tank farms, tankers, and oil fields; never that of the political and military machinations (even more carefully concealed) required to keep the oil (or coal or uranium) flowing; *never that of the mother and father working to pay for their part in all of this.*

Certainly never that of them reproducing their culture in their children by insisting, at the door of the room, that *we close the door when the furnace is on.*

Evidently Comfort is deeply political; and beyond the saving achieved by using a wood-burning stove instead of an oil-burning furnace, this is what is acknowledged by a gesture which overtly and self-consciously manifests the extensive labor otherwise obscured: cutting wood in the backyard for a wood stove is as much a political act as exercise. Here, where the house is heated with oil, the parallel is to Denis and Ingrid's refusal to install air conditioning for the summer and their insistence in the winter on keeping the thermostat at 62° during the day and 58° at night. But whereas the latter insistence shows up in a taste for wearing sweaters indoors, it is the former refusal that causes the mania for opening and closing windows and doors. There is doubtless in this no minor chord of puritanical austerity (Denis in particular is holier than anybody), but it is immediately complicated by a kind of hedonic commitment to *fresh air,* to air not exclusively theirs, to the smells of wet earth, of honeysuckle and privet, to the sounds of bird and beast and other people (to the sounds of their children playing outside), to the call of a neighbor, to the sounds of the city, to the sort of complex sensation of hearing a basketball dribbled down the sidewalk by a kid heading home for the night on a warm summer evening scented with linden blossom.

None of this, of course, is to deny Comfort. It is to complicate it, to make it real, to insist on the inner contradictions that render it slightly bitter even at its sweetest, to insist that it is not something one can have, that it is only something one can be.

By opening and closing doors.

The Doorframe

When the door is open in the spring and nothing fills the space between the jambs but the gossamer of fiberglass, then the doorframe in all its frilly whiteness is close to lace. It is not a fine lace (it is also a wall), and perhaps it is no more than macramé, but there are times, especially from within, when the welling light has dematerialized jambs and transom bar, and it seems no less than a *mouchoir de point de Venise à réseau.*

It is a contrivance of many pieces: it is easy to lose count, but there are forty-two muntins in the sidelights alone, and more than a hundred other jambs, stiles, panels, and pieces of molding. Assembled, these comprise a doorway flanked by rectangular sidelights. Since the tops of these sidelights do not reach to the tops of the light in the door, and since their sills fall a corresponding distance below its bottom, there is about the composition the intimation of a Palladian origin, however distant. Plain doorjambs—which form mullions between the door and each sidelight—terminate in molded plinth blocks. The sidelights are filled with glass set as three tall narrow lights in mitered muntin moldings which dissolve at both ends into fields of quarrels. The space below is embellished with two shallow panels. Doorway and side-

lights together are crowned by a transom—glazed as an inverted fan, flanked by demifanlights—glazed as fans. This *ensemble* wants to be taken for a fanlight, and the outer jambs have been carefully thickened where they curve beneath the crown molding at the ceiling, all but succeeding in creating the illusion that this is something other than a provincial builder's interpretation of a Georgian Revival entrance.

Whatever else it may be, it is the largest thing in the room. Seven feet wide and almost nine feet tall, it takes up sixty-three square feet of all but transparent, and hence lively, wall: in the morning a theater of movement in greys and greens, it is in the afternoon a refulgence of light which all but evaporates the pretense to wall; a frame of shifting chromaticisms in the evening, after the light has faded it becomes a splintered mirror blackened by the night. At any time it is an extravagance of views, each pane giving its own; and what from the door of the kitchen is a single field of vision becomes up close an impossibility of choices. One stands there on winter evenings watching the taillights of the commuters to Cary wink out as they pass down the hill, first through one pane . . . then another.

rules   24. "Keep your hands off the woodwork" (Denis, ktAPP).
25. "Don't kick the doorframe" (Randall, kAPP).

What is the woodwork that one should keep one's hands off it? It is, together with the glass (4, 5), the embodiment of the doorframe in the room. And what is the doorframe that its embodiment should not be touched?

It is lace.

The hole in the wall that is seven feet by nine is filled with lace: the doorframe is a lace curtain (it is a lace-curtain home). This is the meaning of the doorframe, that the inhabitants of the house behind it are lace-curtain people, are people of social and economic standing, are fashionable (they are not shanty Irish). If there are in this analysis traits of a parvenu ostentation, this is in keeping with the origin of the house, meant, like its twins in the neighborhood, to attract the parvenus of Raleigh's growing middle class (people like Denis's grandfather), people who sought in the new subdivisions platted just before the First World War the status and identification that a new, exclusive (white) suburb like Boylan Heights could promise. This status was implicit in covenants controlling the costs of the houses to be built, in the exclusion of blacks from residence, in the curving tree-lined streets, in the proximity to streetcars and parks, in the associations with established wealth and power (Boylan Heights was carved from the old Boylan plantation, and the antebellum plantation house is just up the street); but it was also manifested in the neo-Colonial, neo-Classical, neo-Adamesque, Georgian Revival touches with which the builders cloaked their more vernacular floor plans—the pillars and pilasters, the pediments and porticos, the pavilions and Palladian windows, the *lace* that bespoke . . . social and economic standing, that bespoke . . . fashion.

The equation here was not direct. It was not: lace = wealth and power (though the greater expense of a doorframe of a hundred and fifty pieces could not be ignored), but lace = Culture = wealth and power, where the pretension was not solely to Culture, but also to the modesty that forbade the naked flaunting of wealth and power. This was especially appropriate here where there was no wealth *or* power, where it was no more than the mock wealth and the mock power of the clerks and the floor walkers, of the department managers and the university teachers. The lace of the doorframe was a relay, it flipped a switch that energized the motor of Taste, even here where the fanlight is not a fanlight (where it is a transom glazed as an inverted fan flanked by demifanlights), where the lace is . . . not quite lace, where it is no more than the intimation of lace, where the rough builder's muntins in the sidelights are patently *not* the bars of lead in the windows to which they pretend.

Even here it says: Culture. Even in these early twentieth-century pastiches of revivalist forms a voice with an English accent can be heard muttering about the civil pleasures of a sixteenth-century Italy, these themselves no more than imaginative reminiscences of Rome and a still more ancient Greece. The doorframe *speaks* English (it is an assertion of a certain racial purity) but it is *uttering* a grander genealogy, establishing the owner's origins not in the Amerindian past of a pre-Columbian Iroquoian, nor in the Yoruba of a pre-slaving Africa, nor yet among the oaks of the Celts or back through a Teutonic ancestry to some site in Central Asia, but as the history books in school have it, down the mainstream of Western history, back past England through the Italian Renaissance to the Grandeur that was Rome and the Glory that was Greece. All this the doorframe claims: is there any wonder the kids should keep their hands off it?

But did Denis and Ingrid hear these claims? When they bought the house in which this is the frame for the front door, Boylan Heights had long since ceased being home to the rising middle classes (who now lived in Cary, who drove *through* the neighborhood, their taillights winking out as they passed down the hill). Boylan Heights was: *declining* (it was the decline of the Roman Empire). Denis and Ingrid, intellectual parvenus, were attracted to the neighborhood for precisely this reason: that it wasn't chic. That it was older (they may have been parvenus, but they weren't going to be pushy about it), that the houses had: character (but what did this mean?). That it was close to the university and the downtown (they owned no car). That the house was cheap (as it was they had to borrow the down payment from Ingrid's father). That it was available (they had been evicted from their apartment and had little time to find a house). Given these circumstances, *could* they have heard the claims of the doorframe even had it shouted (had the fanlight been a real fanlight, had the sidelights been set in lead)?

Unavoidably. At sixty-three square feet, the doorframe is the most significant piercing of the wall *in the house.* It was far and away the largest thing in the all but empty room in which it was first appreciated—even in lime sherbet green—as a font of light. Its clumsily delicate tracery of muntins could scarcely be overlooked. Morever, Denis and Ingrid had moved but the year before from a similar neighborhood in Worcester, Massachusetts, where the houses that had most attracted them were Greek Revival; and if this was far from that, there remained nonetheless a relationship. The doorframe was a selling point.

But of what?

Certainly the doorframe could not be for them the racist sign that it would have been for Denis's grandfather (who in the early 1920s had lived in a less pretentious house down the hill): after all, Denis and Ingrid will periodically display on their mantle Nigerian dolls or a Bamboma-Mussuronga statuette, on their record cabinets a Cherokee basket (24). Nor could it well serve as a sign of Georgian sophistication: for from the porch the most obvious thing one will see through the door is a brightly painted, wooden, four-foot-high, Mexican toy Ferris wheel (25). Renaissance overtones will be contradicted by the paintings on the wall: these will be abstractions, primitives (26, 31, 32, 33, 37, 40, 41, 64). Were Rome or Greece to be directly represented in the room, it would be *Etruscan* Rome, it would be the *Bronze Age* Cyclades (it would be a marble of Paros, of Naxos). Unless the confrontation is the *point* (and the room is a debate or a polyphony).

What light do the rules shed on any of this? In the first place, *keep your hands off the woodwork* doesn't mean what it says. What it means is, *don't get your dirty hands all over the woodwork,* that is, don't get the woodwork dirty; and, *you don't write on the woodwork or anything else,* where the *anything else* reflects the afterthought that there are other things the kids could do to the woodwork which may as well be covered, however vaguely. At issue here is the *marking* of the doorframe, or *marring* it as in RULE 25 (though kicking the doorframe *chips* the white enamel and exposes the lime sherbet green—that is, *marks* the doorframe). There are two objections to marking the doorframe. One is that getting it dirty annuls any claim the *doorframe might otherwise make* to participation in the broader significations of its form (dirty lace = shanty Irish = no Culture): *if you're going to get it dirty you might as well not have it.* It is called *wood*work, but all signs of wood—of tree—have been obliterated: trees have been hewn, sawn, milled into scantling; from these a doorframe has been assembled; this has been painted a glossy white. Despite the *denomination* of nature, there is little of nature here: it is, literally, cultural (it is a highly finished artifact). To dirty this is to return it to nature, to undo what human culture has accomplished, what painting it white confirms.

The other objection is that marking the doorframe (that is, getting it dirty) implies a double abandon *on the part of the marker* (the barbarian, the kid), neither of which can be accepted within the house (within the domain of culture). The first of these is simply getting dirty: this is always of an animal order. The second is a wantonness close to that forbidden with the slamming of the door (RULE 3): *what were your hands doing all over the woodwork anyhow?* There is inevitably in this question the imputation of a sensuous fondling, a rubbing, a caressing of the doorframe that is also of an animal order (it is in any case: *Dionysian*): *what were you doing?! My God! Look what you've done to the door!!* For the rule-maker this is the interminable struggle: to *preserve* the culture (the doorframe) against the onslaught of nature (young kids), while simultaneously—through the medium of the preserving rules—*reproducing* in the kid the culture (so that the rule-maker can relax). But what culture is this?

It is classical: it is precisely that embodied *in the form* of the door.

In some sense culture is inevitably classical, for the transmission logics exploited in the reproduction of the culture are deeply conservative; and in this sense it is all but tautological to assert the classicism of the culture carried by the rules. But in fact, the culture in question is more than classical, it is Apollonian. Against the sensuous, frenzied *(don't kick the doorframe),* orgiastic, irrational *(what are you doing?),* unbounded, lawless behavior signified in the kicking, in the marking, in the dirtying of the doorframe, the rules oppose a measured, restrained, temperate behavior of a rational, nomothetic character. What is this if not the very image of the Greece that is echoed in the form of the doorframe, in the choice of the white in which to paint it (in the pillars on the porch, in the restraint of the numbers that identify the house on the outside of the doorframe). That none of this has anything to do with Greece (which was also Dionysian, which did *not* have white temples) is irrelevant, *for it has everything to do with the Greece that is reached in the mainstream of Western history as the font of the culture that the rules strain to reproduce.*

It doesn't matter whether the iconography of the doorframe is understood: it is sufficient to maintain the doorframe in the best possible condition to act out all the implications of its form. It is as though the culture did not trust itself to the artifact or the act, but from the construction of the former through the latter, embedded itself with equal pertinence in both. Yet in obeying the rules with respect to this *doorframe,* Randall and Chandler will obey the rules with respect to this *form,* neither of which— this must be insisted upon—lacks reinforcing animal pleasures (the crafted panes of the quarrels, the exuberant light, the hierarchic subordination of the panes of glass, the simple symmetries). In time, the pleasure in the light and the sense of balance, the form of the doorframe and the form of the rules, will become . . . inextricably enmeshed: barbarians will have been tamed, the kids will have grown up.

The Window in the Door

To lie on the couch and look out the window in the door is to look up and out to the gradually dying pin oak across the corner—clouds, clouds of leaves, but also branches every year more bare. Toward the bottom of the window the roof of the house beneath the tree obtrudes, but at the bottom of the window leafy green again asserts itself in a strip as long and deep as the bevel in the glass. This, spanning more than an inch, is cut around the perimeter of the pane—the inside is flat—though almost half of it is hidden by the mitered stool molding employed to secure the glass in the opening framed by the rails and stiles of the door (2). The glass is plate (not window)—that is, rolled (not drawn)—and, at twenty-six by forty inches, almost seven square feet. It's a generous window, for a door that closes with a *thunk*.

Afternoons, when the light is right, the bevels cast rainbows on the walls.

rules   26. "Keep your fingers *off* the windows" (Denis, ktAPP).
27. "Don't smudge up the windows" (Randall, ktAPP).
28. "Don't breathe on it" (Chandler, ktAPP).
29. "Don't do anything that will break the glass" (Ingrid, ktPRO).

What is the glass that one should not even breathe on it?

It is a gem.

This is the meaning of the bevel: that the window is a crystal, that it is a gemstone, that it is a table-cut diamond (the doorway is a throat with a lace collar and a diamond brooch). Since the bevel on the glass is a more subtle signifier than the muntins of the doorframe (it is less easy to see), it has a correspondingly higher threshold of significance. This is not merely because the glass is easier to wound than the doorframe (so that its preservation more surely bespeaks a well-ordered life), but because it displays in an even more outrageous fashion than do dirty fingerprints on the glossy white woodwork of the doorframe the signs of its transgression. Whereas the signs of the window's *contrivance* are all but lost in the transparency of the glass (all traces of tank car loads of silica, alumina, lime, magnesia, soda, all evidence of furnaces, high temperatures, rollers, annealing ovens, grinders, and polishing machines have disappeared), the signs of *keeping it clean* are all but inescapable. Windex, which has "Tough Grease Cutting Power," cleans glass with Ammonia-D. Next to the word "Shines" on the label is the image of a light flare: the fascination with the diamond is a fascination

with the stars, it is a star fallen to earth (it appertains to the heavens, the window is heavenly). Cascade "with advanced sheeting action" now "Fights Spots Better than Ever!" The light flare on the rim of a crystal goblet reads like a brilliant on a ring. Spic and Span ("DANGER: In case of eye contact, flush with water for 15 minutes. Get prompt medical attention.") spells the equation out: "So clean it shines"—that is, so shiny it must be clean, where shine = clean (= decent). Thus while the bevel draws attention to the economic status of the householder (lace may be nice, but diamonds are forever), the cleanliness of the glass draws attention to the moral qualities asserted by the dictum that cleanliness is next to godliness (as the light flares from the glass on the label recall us to the heaven that is god's house).

The touch of holiness is balm for the sting of the pretension to wealth.

There is in this concern for appearance no little admixture of a concern for control. After all, the work of the window is to transmit light, and it does its work less well when it is less than clean. The alibi of a simple pragmatics lurks ever ready to deny the stars, the diamonds, and the morals of cleanliness: the cleaning is a matter of efficiency, of daylighting. And doubtless this is true, doubtless the window does serve this function, and beyond the pragmatics of lighting, a joy in light as well. But no one entering the room through the door can possibly miss, off to the left, a three-quarter-inch thick slab of glass eleven square feet in area, with bevels cut on top and bottom as well as at the corners (65). No doubt this table can claim its functions as well, but they are of the class of the Orefors wine glasses that sometimes sit upon it—they hold wine, and it holds them, but with an excess that is the very definition of their attraction. *Not breathe on it?* And even when it is accepted that this refers to the wintertime when warm moist breath leaves a mist of condensate in which lines can be traced that leave the tracks of their trail long after the condensate has evaporated, still the rule in *its* excess casts into relief the role of the glass and its cleanliness in the social and moral economies of the entrance.

The Glass in the Sidelights

In the late afternoon of a midsummer's day the sun spanks the north side of the muntins in the doorframe, knocking their inner ends into deep local shade. This makes moldings of light, dim and bright, that frame a lambence of sun on leaves that from a certain angle is all that can be seen through the glass of the sidelights. There is this dark edge, a fillet of light, and a play of flaxen viridescence, the dark edge, the fillet, the play . . .

Depending on how you count, this happens thirty-four times, for twenty demi-

quarrels, eight quarrels, and six long skinny hexagons flank the doorway in a mesh of mitered muntins—there are forty-two of these—supported in their frame of rails and stiles (3). Given the portion hidden behind the molding and the glazing compound (applied to the outside of the windows), just five square feet of glass is free to project on the living room floor its cabalism of triangles, squares, and hexagons.

rules
30. "Don't breathe on them" (Chandler, ktᴀᴘᴘ).
31. "Don't touch the windows" (Randall, ktᴀᴘᴘ).
32. "Don't let your friends put their noses on the windows" (Denis, ktᴀᴘᴘ).
33. "Don't wipe off the windows with your hands. You only smear them up" (Denis, ktᴀᴘᴘ).
34. "Don't pound on them" (Chandler, kᴀᴘᴘ, ktᴘʀᴏ).
35. "Don't spit on them" (Chandler, kᴄᴏɴ).

If the glass in the door is a diamond by virtue of its bevels, the panes in the sidelights are diamonds by virtue of their shapes (the eight full quarrels are literally diamonds). The doorframe may be lace, but the glass in the doorframe is a graduated *rivière,* the doorway *is* a throat, and its passage signifies a swallowing by the home (when the guy who mistook the house for his friend's was shown the door, the home coughed politely: it *can* throw up). Given the exclusively signifying function of a necklace, there is no point in attempting any case for a pragmatics of the glass in the sidelights: they are fully saturated by significations entirely social, economic. Consequently there is not the slightest pretense to utility: the smallest panes are just over three square inches, have two angles of 45°, and are impossible to wash. Even the full quarrels are tiny, and the elongated hexagons are not more easy to clean. Yet the importance of keeping them clean increases as the difficulty, and whatever delight the glass may give (and it is considerable) remains wholly attendant on its cleanliness. Given the sizes and shapes involved, this is easier to preserve than attain, and hence the elaboration of the rules articulated to assure the appearance of the window in the door to hysterical levels of specificity: *don't let your friends put their noses on the window.* Yet Chandler noted in his comments about the rules that "Frank, Billy, and William do that" (where, again, "that" refers to their putting their nose—or lips or forehead—on the steamed up windows of the wintertime); and it is Chandler who emits the rule protecting the panes from pounding, reflecting an awareness (however inculcated) of the difficulty surrounding their replacement. What might have been a necessity when sheets of glass were limited in size by the difficulties of their manufacture is, in an age when sheets of glass are produced in unbroken streams, the sign of a luxury. Though it is not the luxury that produces the pleasure in the panes (this is attributable to the play of light, to the multiplicity of views), it is the luxury that is reflected in the rules.

To Chandler's *don't spit on them,* Randall added, *on any window.* We have cited this in the control code out of deference to the covering rule, *we keep our mouth water in our mouth* (a rule inherited through the kids' tenure in nursery school). There is also reason to believe the rule an artifact of our collection, since the adults don't recognize it and the kids can give no history.

The Bells on the Door

There comes to deep in the house a clatter as of metal plates falling to the floor and a high hulloo: someone's come home.

This is not the doorbell. The doorbell, activated by an illuminated button set into the exterior casing of the doorjamb, is an electric chime—actually a two-bar metallophone with an individual resonator for each bar—located in the dining room. The clatter is from other bells, from a rope of three Indian camel bells (suspended from the neck of the interior doorknob) set jangling by the initial push needed to start the door: because the bells are lighter than the door, they are propelled forward faster than the door, so that when the door catches them up, it does so with a *clitter-clatter,* with a *rattlety-bang* (without a sonorous tintinabulation). Much of this is caused by the bells' bodies whacking into the lock stile of the door, though clappers surviving in two of the bells add a tingling and a jingling to the rest of the racket. The bells, graduated in size (a higher-pitched fourth bell is missing), are constructed of sheets of a copper alloy bent into riveted cylinders. To crowns soldered to one end of the cylinders have been affixed rings whose ends have been bent within the bells to support similarly constructed clappers. The rings have been knotted onto a sisal braid in black, gray, and orange. There is something barbaric about the whole, a kind of peasant crudeness (it was manufactured for export consumption) that sharply contradicts the burgher pretensions of the small glass panes, the bevels, the lace, and the Georgian Revival references, something which can also be seen in its history, a string of bells purchased thirty years ago in Greenwich Village by Denis's parents from a little shop on 8th Street when Greenwich Village was rather more *épater le bourgeois* than it has become today, when the simple camel bells had some of that éclat.

When the door closes, it is not thus merely the high-pitched sound of the latch scraping across the tongue of the strike plate and the lower sound of the door falling against its stop. It is also the brattle of three idiophones brought here from the antipodes stamping and hitting and being shaken and being hit. And when the door is opened, there is often also a high hulloo.

rules   36. "Don't play with the bells" (Denis, tcon).

37. "Don't dingdong them too much when someone's sitting in the living room" (Chandler, ktAPP).

38. "Keep your friends from dingdonging them too much" (Randall, ktAPP, tCON).

Why are there two bells at the door?

Because there are two classes of persons who use it. On the one hand, according to RULES 6, 7, 17, 18 and 19, there are those for whom the doors are not to be opened by Randall and Chandler; on the other, there is a class of persons implied by RULES 8 and 9 (inclusive of Randall and Chandler) who may open the doors at will. That is, there are strangers and certain others, who in order to gain entrance are obligated to ring the doorbell, and guests, friends, and family, for whom ringing the doorbell would constitute at very best a peculiar formality. These latter open the door—*clitter-clatter, rattlety-bang*—and come in.

In this manner the two forms of bells concretize aspects of the taxonomy of persons established by the door: *they give them voices,* the dulcet tones of the doorbell torn between announcing strangers and trying to pretend it can only be heard in the servants' quarters, and the din of the camel bells chattering at the entrance of those too familiar to ring. But the bells do not simply signal the difference *strangers/friends,* but *at the door/not at the door,* a distinction established sonically by the sound of the bell emerging out of silence. When this silence is broken by the sound of bells independent of the variant *at the door/not at the door* (and Denis or Ingrid enters the living room only to find the kids playing with the bells), the rule *don't play with the bells* is emitted with greater or lesser force as a way of maintaining the environment necessary to permit the bells to mark a difference. This is a simple matter of control akin to that of *close the door every time you go through* (RULE 4) or *open the door for fresh air* (RULE 23): its end is to enable the bells to *work,* that is, not *play*—where play is of precisely the order of animal abandon (*just ringing the bells*) as that indicted in smudged windows and dirty woodwork, that is, where the explicitly human and cultural is betrayed by the uncivilized barbarian. In this way issues of control are contaminated by issues of appearance, control is forced to forsake its alibi of utility and revel in the reality that the work done by the bells is superfluous, is *play,* is of the same order as that done by the bevels in the glass, by the lace of the woodwork. This is attested to historically, for the bells appeared on the door only long after not just the doorbell but the door latch itself had ceased working; that is, when anyone knocking on the door (to signal a presence) opened the door (through the force of the knock)—that's when the bells appeared, as a device to notify the occupants of the house of the violation of its membrane (as in, once again, the story of the guy looking for his friend). Two things must be observed here. One is that this entailed no restructuring of the classes of persons using the door: those who now ring, knocked; and those who come on in, came on in hulloing with the camels (the voices were different but the structure remained the same). The other is that the alibi of the security function holds up only so long as we ignore the fact that neither side nor back door have ever been equipped with either variety of bell. Yet the traffic at the back door is no less than at the front (though it is differently composed: for example, the one thief to have violated the house entered through the back door). Thus it is not clear that the bells at the door ever had any *work* to do, though had they, this had long since been completed (the door latch and the doorbell have been repaired). What then are the bells doing on the door and why may they not be played with?

As to the latter, playing with the bells is doubtless forbidden on the same grounds as slamming the door (RULE 3), less because of its impact on the environment in which a sounding bell can signify (RULE 36), or even its irritation (though RULES 37 and 38 obviously derive their surface authority in this way), but because its licentious wallowing in wide-open sound offends the sense of the Well-Ordered Life so evident in the unspotted woodwork and clean windows. However, the emphasis here needs to be on the sign functions of *unspotted* and *clean,* for there is no doubt that the bells are encouraged to oppose themselves to the doorframe (and its allies) in *their* sign function. Here the twisted braid counterpoises itself to the smooth, glossy carpenteredness of the woodwork, its orange-gray-blackness to the whiteness of the lace, the crudity of its craftsmanship to the machined slickness of the glass, its raucousness to the mellow chimes of the doorbell. Without overlooking its security function (not the less real for being incomplete), or ignoring the perceptual pleasures to be derived from the soniferous complications enriching the aural culture of the room, something else is introduced by the presence on the door of these Indian bells, an iconoclasm the more powerful for its peripheralness: this is not folk art on a pedestal, but folk art marginally warped into the utilitarianism of the room, hanging from the knob of a door which is the very sign of everything a camel herder's life is not, the mark of a domesticity the nomad (at least *as sign*) is precluded from knowing. The bells are the sign of a doubt echoed everywhere in the room, in the wicker rocker (35) and the conversations launched from it; in the painting of the Italian anarchist Malatesta (41) and the conversations held about it; in the wooden Mexican Ferris wheel (25) and the conversations it inevitably starts. It is a doubt which, while ultimately recuperable (the rocker was a bargain, the painting is by Denis's brother, the Ferris wheel is a souvenir), is nevertheless permitted to permeate the room. The bells are the support, among others, of an unease, of a question, of a hesitation: rules, as of the doorframe (RULES 24 and 25), which were seen to converge with its form toward an image of Western classicism, will be asked again and again to converge with other forms toward other images of human life: the room will acquire a diffuseness: for all the whiteness and lightness and space, the room is pervaded by a haze of signs, by a failure of determinism, by a buzz of mental complications, by an . . . effervescence of meanings.

X · THE VOCABULARY

Where do these meanings come from? Not, certainly, from the objects themselves. In themselves the camel bells are scarcely bells, certainly they are not camel bells let alone Indian camel bells. On a camel, in India, they would never be curiosities, their "folk" attributes would be invisible, even their crudity would be less apparent. To become "folk art" they would need to be placed in a matrix of opposition to "popular art" and "high art"; to turn into the mark of a hesitation they would have to be hung on a glossy white door with Georgian Revival touches in a room in which other objects of similar character were displayed in a manner that established *their* credentials as art. Meaning leaps from an opposition between what is instantiated and what is not (it is a braid of Indian camel bells that hangs from the doorknob, not the cord of finely wrought Indian brass claw bells that hung from a bracket in Denis's father's sister's home). For the bells to sound their meaning (though this is as likely to be visual or tactile), an auditor needs to know the sounds of the bells that are *not* ringing; and if this is metaphor, at the level where the bells act as a signal it also happens literally to be true.

And they *are* a signal. Whatever other functions support their presence on the door, this one cannot be denied: that when the rules are in effect, the sound of the camel bells signals the opening of the door. We have seen that it does this first by opposing itself to silence *(sound/silence)*, but that, on the next level of meaning, it opposes itself to the sound of other bells as well *(camel bells/doorbell)*. For those living the room this opposition is but one of a sequence of oppositions that knits up into a mesh the bells at the door with those elsewhere in the house, in the world—out in the kitchen and up in the bedrooms, stuck away with the Christmas things or out on the back porch, or *off* the porch on the backs of backing trucks or the walls of buildings being burglarized, in the halls of schools or in the homes of others, on old cash registers or in orchestras and bands. The ability the camel bells have to play the role they do in the structure of the door is attributable to the positions they occupy vis-à-vis these other bells; for the camel bells are not crude, let us say, absolutely, but only by virtue of the differences that distinguish them from the fine silver handbell—from Ingrid's mother's family—used by the sick to summon aid and comfort (how Ingrid's mother's family also used it), but which otherwise sits in a china cupboard reflecting its antecedents as a bell for the maid to bring on the next course or clear the table. Nor is this bell fine in any absolute sense, but only through its oppositions to the bell Denis keeps in his bedside basket to use when he is sick, a cheap thing of pressed metal—barely campaniform—with a blue plastic handle in the middle of swirly stenciled stripes of red and green and yellow blobbed with dots and stars, a Japanese tin toy from Goodwill that Ingrid found one year to put in Denis's Christmas stocking . . .

The three bells sketch the paradigm *folk/elite/popular,* but the completed domestic campanology sketches the world, reproduces it in all its vital oppositions. Thus we find the sequence *machine/man/nature* reproduced when we move from the *synthesized beeps* of the kitchen timer, the kids' watches, Denis's office and travel clocks, and the smoke detector; through the *clangor* of the old wood-handled brass handbell that calls people to dinner; to the *plangent ripple* of the windchime hanging from the eave on the back porch (in which sequence electrically driven nonsynthesized bells—as in the clock in Denis and Ingrid's bedroom—constitute a glissade term between machine and man: *machine/——/man/nature*). Or we observe the pair *sacred/profane* in the distinction between the Christmas bells (the tiny brass Indian and little American steel claw bells of the Christmas stockings, and a porcelain handbell Randall won as a Christmas prize in the fourth grade) and the telephone bells (themselves stretched between *old fashioned/modern* as synthesizers replace electromagnetic clappers). Or we note the opposition *work/play* between the doorbells, telephone bells, the bell on the stove (now defunct), and those in Chandler's electronic "arcade" game, those in his and Randall's old crib toys, those on certain pull toys and rhythm instruments, the three sleigh bells on their strap of leather, and the clay bells from Coyotopec, in Oaxaca, in Mexico. The bells vary in pitch *(high/low),* tone *(clear/clunky),* volume *(loud/quiet),* consonance *(ordered/confused* or *tuned/untuned),* and craft *(crude/fine),* invoke an entire economy of substances *(wood/low-fired clay/high-fired clay/copper alloy/brass/tinned steel/stainless steel/silver),* and come from ten different countries (India, Germany, Japan, Mexico, Yugoslavia, Hong Kong, Taiwan, Greece, England, the United States). It is in its position in the multidimensioned matrix generated by all these scales (and others) that each bell finds its meaning, or meanings, for the unitary matrix is never more than the product of an analysis, and meanings radiate from each bell depending on the path chosen through the bell-meaning space, from beeper to alarm clock to school bell, for instance, or from the silver bell to the windchimes to the cascading tones of the carillon behind the "Te Deum Laudamus" in *The Play of Daniel* (a reference the less arcane when it is recognized that the old Noah Greenberg recording constitutes the cornerstone of the home's collection of Christmas records [27]). So the paths wander, but not at random—but so that if walked long enough the whole world could be taken in from any point of origin.

Though it is the *things* in the room (in the home) that are the objects of the rules (the subjects are the kids, the barbarians: *don't [you] play with the bells*), it is along the paths charted through their mutual affinities and oppositions *that the meanings are manifested toward whose conservation and reproduction the room/objects/significance-thing is directed.* Insofar as this is directed toward a Well-Ordered Life, and insofar as this direction involves control over the membrane of the house, and insofar as this may demand a signal to mark its violations, and insofar as these may be appropriate (guests, friends, family) or inappropriate (strangers, certain others), then the *clitter-clatter* of the camel bells will signify along the route *silence/camel bells/doorbell*—while nonetheless retaining the ability to signify along any other route open to it in the matrix of affinities and oppositions (for example, along the route *folk/popular/elite*). Rules, however, will be articulated only with respect to threatened meanings. They will thus act to control the use of the bells to preserve the silence necessary for the bells to signify in their signal function (along the route *silence/camel bells/doorbell*); but inasmuch as their folk art function is either not threatened by the kids, or is adequately covered by the effect of the rules articulated in the control code, no rule will be explicitly evolved to deal with it. Should a heretofore unthreatened meaning become so—*guys! you know better than that, we always ⸺!*

This is not to say, however, that such a meaning is not embraced by a code, for the codes alone secure the meanings (codes are abstractions of the structure of the matrices of affinities and oppositions). While we may interpret the explicit rules *(don't play with the bell)* through reference to certain codes, these codes are not exhausted by the explicit rules: all behavior is given significance through the codes, whether or not caught up in the net of words. Thus although there is no positive rule for the use of the camel bells, the prohibition of their being played with insures their ability to signal the opening of the door, that is, acquire meaning through a subset of the control code that specifies *camel bells ringing = opening door;* and doubtless this positive function was many times expressed in the attempt to justify the rule in its transmission (but . . . perhaps not). This strict codification enables the bells to work (*or* play):

Ding-dong: and the footsteps of someone going to answer the door.

Clitter-clatter, rattlety-bang: and a high "Yoo-hoo."

Clang, clang, clang: and footsteps coming down the stairs for dinner.

Beep . . . beep . . . beep: and the sound of the oven door being opened.

Tinkle, tinkle, tinkle: and after a moment (while Ingrid dries her hands) the sounds of her feet upon the steps going up to see what her sick son wants.

How are the *things* different from *words*?

Are not these bells like those in that list in the thesaurus—*bell, tintinabulum, gong, triangle*—that falls within the purview of "Specific Sounds" under "Hearing" in the class "Sensation"? Do not the things of the room (of the house, of the world) comprise a lexicon? Is it not out of its units that we assemble—in accordance with the grammar that is the living *in abstracto*—sentences like "door"? (Not: *door proper* or *screen door* or *bells on the door* or *doorframe* or *sidelights* but . . . "door"?)

And is not the door we have been trying to get through (the door to the room), is it not such a sentence? Is it not more than scantling and screening and brass and glass but a powerful syntagmatic structure whose final meaning depends, not only on its syntax, but on the systematic relationships among the units from which it is constructed? That is, does our assertion that the beveled glass of the door's window is a diamond rest, finally, on no more than simple resemblance? Or does it depend on the existence of a mesh—a network—of oppositions (and affinities) knitting up at least the glass of the house (inevitably embedded in that of the world)? Consider only the instances of *beveled* glass in the house. There are eleven of these: the window in the front door (4); the table in the living room (65); the window in the door from the front porch to Denis's office (often called the "side door"); the mirror in the overmantle of this room; two bowls and a plate of crystal, etched and cut, and stored in the china cupboards in the kitchen above the silver bell; and a crystal salt shaker and three crystal bud vases kept in a milk cupboard in the dining room—the crystal all from Ingrid's mother, or through her mother from her side of the family. This *affinitive* glass connects the window in the living room door to: the front of the house (the face), the living room (the parlor, that is, conversation, guests—where conversation is doubly related to the front of the house through the mouth), the scholar's office (social status, cerebral activity, more conversation), elaborate mantles (again to the living room, to the face), dining room (and so to guests, to eating as ritual), china-silver-crystal (eating as ritual, guests, social status, Orefors, wine, sophistication, conversation), floral buds (roses), vases (decorative superfluity, social pretension, what Maggie always threw at Jiggs in *Bringing Up Father*). The sequence of *oppositions* leads through the *unbeveled* glass, a class almost too large to deal with here, comprising the balance of the glazing in the windows and picture frames (5, 17, 31, 32, 33, 37, 47, 49, 50, 52, and 56 in the living room alone, to say nothing of the window in the back door), the light bulbs (34, 42 in the living room alone), the mirrors (six in the bathrooms alone), the "regular" drinking glasses (cheap dime store water glasses from Goodwill), the dessert plates and bowls, the Pyrex measuring cup and other glassware, the old milk bottles (used as pitchers), and the glass jars and bottles of various sizes, both filled as they came from the store, and as reused (for example, to root geraniums in), or waiting to be reused, a large stock, in the cupboards above the refrigerator. This oppositional glass separates the window in the living room door from: the back of the house (the behind, the anus), the kitchen (opposed to the living room–dining room as cooking to eating, and so as labor to leisure, as worker to owner), the stove (opposed to the decorative mantle above the inoperable fireplace), the bathroom (doubly related to the back of the house through the toilet), milk (as opposed to wine, so children as opposed to adults, provincial as to sophisticated), rooting geraniums (opposed to dead cut flowers, opposed to show roses). The opposition between cut and uncut glass resolves itself as an opposition between the face (front, mouth, conversation—we had previously seen that the doorway was a throat) and the rest of the body (specifically the muscles—labor in the kitchen—and anus—trash, garbage, feces, urine).

Why is this path through the glass-meaning space privileged? It is not. There is no privileged path, there are as many paths as there are forks, and there are as many forks as paradigms in which the glass participates. We observe, for instance, in the distinctions between the lights and balls of Christmas—ceremoniously taken from their tissue-lined boxes each December—and the Pyrex food containers taken from the ice box and popped into the oven, the opposition *sacred/profane;* as in that between the pressed-glass plate ("Give Us This Day Our Daily Bread") and the slick salad bowl, the opposition *old fashioned/modern.* In the distance between the Orefors stemware and the Duralux bistro tumblers we see something of the distance between *high social status/low social status,* an opposition we note is paralleled by one of substance *(high lead oxide content/low lead oxide content).* The glass varies in point of origin *(Sweden/Germany/France/United States),* size *(large/small),* form *(rectangular/circular* and *flat/curved),* surface *(etched/unetched* or *cut/pressed/smooth),* age *(old/new),* and provenance *(of the house/of the Family Wood)*—to note just the first that come to view—and were we, for example, to take the route suggested by this last pair, the affinitive sequence sketched just above for all glass with bevels would be seen to collapse into the oppositional sequence of *door-windows-mirrors* (provenance: of the house)/*cut tableware* (provenance: of the Family Wood); at which point the crystal bowls, plates, shakers, and vases would be released from their association with the window in the door and be freed to enter into novel sequences such as *Ingrid's side of the family/Denis's side of the family,* an opposition particularly relevant to both the glass—for as we have seen, *all* the cut glass comes from or through Ingrid's mother—and the living room, for as we shall see, *nothing* in the living room comes from Ingrid's side of the family.

Then what *does* come from Ingrid's side of the family? Precisely those things most deeply immersed in the everyday rituals of bourgeois life, especially, centrally, eating: that is, crystal bowls and gold-rimmed china, silver forks and spoons and serving platters, heavy linens, woven table cloths, lace doilies, tea cosies, special cake forks with one tine flattened into a knife, tongs for serving little cakes, even—symptomatically—

silver "straws" for the sipping of tall cooling drinks. That is, things decidedly not of the kitchen—though largely stored there—but . . . of the dining room. Contrastingly, almost nothing of the dining room comes from Denis's side of the family, and when it does, it is always disguised: as folk art (it is a large, fragile, low-fired, green-glazed Patamban platter).

It is interesting: there are extensive connections with Mexico on both sides of the family, manifested equally as artifacts in the home. On Ingrid's side these are all but exclusively heavy Mexican silver (they are: sterling serving bowls, sterling candy dishes, sterling petits fours trays, sterling coffee sets, sterling "straws": they support ritualized eating). On Denis's side they are crude pottery: an all but unfired pottery "drum" (39), a pottery statuette (55), a brightly painted toy Ferris wheel (25), a toy wooden car (60). Almost every cup of coffee drunk in the home is made in Mexican silver brewers: the silver from Ingrid's side of the family is *used,* daily, in eating. A lot of this coffee is drunk in the living room, surrounded by the Mexican folk art that comes through Denis's side of the family, and if most of this was originally his own, certainly the toy wooden car was a gift from his parents, as were the camel bells (6), a lithograph (40), a photograph (56), and many records (27). If an isolable aspect of Ingrid's family life can be synopsized as a celebration of the rituals of eating, then an isolable aspect of Denis's family life can be synopsized as a practice of the arts of contemplation (of music, books, paintings), an activity evidently no less bourgeois in origin and character than the family supper, but one which frequently distances itself from this as distinctly superior ("They have no books!" or "They have nothing on their walls but mirrors!"), as somehow more significant, as even faintly . . . bohemian. Both of these aspects have been reconstructed in the home that Denis and Ingrid live (the home is a form of reconciliation), but as an opposition between the living room (brought forward from Denis's family) and the dining room–kitchen (brought forward from Ingrid's family), an opposition that not only concretizes a certain sexual dimorphism (no matter how much Denis and Ingrid might labor to contradict this), but which precisely reproduces the opposition between the front of the house and its back, the same that we saw underlay the opposition between the beveled and unbeveled glass, namely that between the face and the rest of the body.

Thus a seemingly distant path has brought us back to where we started, to the very paradigm which furthermore is underwritten by each other part of the door except for the screen and the camel bells. It is not that this is so particularly attractive a paradigm that we find it difficult to escape (it is abhorrent), but that it clings to the door with the tenacity of syntax, one moreover guaranteed by a certain logic of the house as a whole. Thus the quarrels in the sidelights are a decorative feature of *all* the windows in the living and dining rooms that are not beveled (17, 47, 50) but of only two of the remaining ten sashes on the ground floor, and of none in the basement or second floor. Here again the front of the house is marked as different from the back, and in fact this common and omnipresent paradigm gushes from the house onto the lot which *just happens* to run from the paved street where the guests park their cars to the unpaved alley where the trash cans await the arrival of the (black) garbagemen (compounding the existent sexual dimorphism with a racial one).

Here, however, it cannot be overlooked that the opposition marked by the spatial distribution of the quarrels is actually *living room–dining room/rest of house,* whereas that reached along the route *Ingrid's side of the family/Denis's side of the family* was expressly *living room/rest of house.* This failure of isomorphism makes it clear that whatever the appearances, the family and the oppositions it embraces do not reproduce those of the house. They infiltrate the house in solidified form: neither house nor family determines the *home* (the room will be neither architecture nor sociology), but through the living achieve *accommodation.* (The home is thus a dual reconciliation, between Denis's and Ingrid's families, and between this reconciled Family Wood and the house in which it dwells.) This is what must be grasped at the door, that the house—vernacular precipitate of the social competence that constructed it—*will be heard.* Let it be accepted that it is the tenacity of the house's syntax that gathers at the door those threads woven in the matrices of glass and wood that insist upon some form of the paradigm *mouth/anus.* Let it further be accepted that this paradigm was more or less weakly embedded in the structures of those families whose memories are lived in the living of this home. At the door, the syntax that unites the signifying units is a free form: it is a relation of simple combination that links a certain number of matrices seen in a certain pertinency into a single utterance: in a Georgian Revival door with beveled glass and quarrels in the sidelights, this can only be *classical patriarchal Western culture supporting and supported by wealth and power.* In the desert of pretension that is this door, that is this proposition, that is this articulation of a nauseating sentiment, that is this statement, this syntagm, this concatenation of meanings, only the camel bells—only the camel bells and a certain awkwardness and a subtle sloth—introduce the vital irritation of a vital living, a *sure, you bet* to the smug proclamations of the door, precisely as, unbelievably enough, the door will introduce to the room the necessary hesitation to its smug counterpretensions of being something other than the manifestation of a bourgeois culture that it is: the door will say, *sure, you bet.*

Inevitably the things of the room *are* the vocabulary out of which its sentences must be formed. Its limitations, its perversities come with its potentials, its rightness, its instinct for the true. It is like any other vocabulary (it is like this vocabulary): in enabling the struggle for speech, it speaks *itself,* its own histories, its own distortions (its own truths). It doesn't just get *in* the way, it *is* the way. And when this way is threatened, when the meaning is about to be obscured—

"C'mon, guys! Let's keep our hands off the woodwork, okay?"

XI · TONE OF VOICE

Sure, you bet.

In what tone of voice is this said? It is said in a tone of sarcasm, it is said mockingly. However we know this (whatever we hear), it is something we can know not only of utterances but also of things. The camel bells mock the pretensions of the door, the door mocks the pretensions of the room.

A system open to the expression of mockery is always a ternary system. Because the possibility exists for discounting any expression, an intensive form evolves to prohibit such a reading, or to contradict it.

Sure, you bet.

No, really!

In the room, certain locations, certain juxtapositions constitute privileged sites for the expression of this form. For example, anything in the center of the mantle is immediately granted the seriousness of its position, a seriousness paradoxically heightened by any proclivity the thing might have toward emptiness, toward mockery. Alone on the mantle, in its center, is a piece of copper wire Ingrid found in the street (63). The opposite of what is expected (it is not a tambour clock, it is not a Chinese vase), it doubles the seriousness of its claim to be taken . . . seriously. There can be no question of the wire's right to be where it is: the syntagm says, "No, really!"

"Isn't that beautiful? Putting it on the mantle really brings out its beauty."

Anything placed on the easel (19) is, through the agency of this gesture, similarly endowed with status. Had it lacked status previously, its placement here would be sufficient to constitute a heterodox assertion of its value. Denis has never let a friend forget his remarking of a *Star Wars* poster placed here, "Oh? Are movie posters to be art now?"

You can hear the eyebrows being raised.

The system at work here is no different from that operating to signify the bells, the beveled glass of the quarrels in the sidelights: it is one of differences set in paradigmatic relationship: *edge/center,* for instance (where the couch, seat of guests, protrudes into the center of the room); or, *underfoot/eye level/overhead* (where the paintings, drawing, lithograph, photograph, assemblage, collages are all at eye level!). The paradigmatic relationships holding among these *things* (their positions in the matrices of oppositions and affinities) work to certify the paradigmatic relationships holding among these *locations*—and vice versa: the paradigms are mutually sustained by their homology. Admittance to the house alone constitutes an ascription of value to the object (precisely as the object flatters the house). Higher values are inscribed through movements toward the ground floor (devaluation is achieved through movements toward attic and basement), then through movements toward the front of the house (according to the paradigms *front/back* and *mouth/anus*). Highest values are accorded objects at eye level in the center of the living room. Actually, this place is vacant (it is a modern room, it participates in modernism's distrust of centrism, of axiality), but it hovers above the right end of the couch in a force field of objets d'art.

Yet to place this couch, these paintings, this table *here* is to speak in a normal tone of voice: it is no more than expected, it is polite, it *lives* the paradigm. To have put here objects of lesser (ostensible) value would have been to speak in an off-tone, among which three principal forms can be distinguished: the exploitative, the assaultive, the evasive. The first exploits the locative paradigm to *invest* the object with increased value (as in Denis's friend's ascription of increased value to the *Star Wars* poster, as in the copper wire on the mantle [63]). This tone of voice not only acknowledges the paradigm, but *puts it to work*. It exploits it. Correspondingly, it is the tone invariably taken by art galleries, museums (it is often cynical). The second exploits the object paradigm to *divest* the location of its significance, to strip it of its pretensions (the canonical example is Duchamp's submission of a urinal to the exhibition organized by the Society of Independent Artists for the Grand Central Galleries, but the camel bells [6] have something of this tone with respect to the door, as has the door with respect to the rest of the room). This tone of voice acknowledges the paradigm, but only *to tear it apart*. It assaults it. Correspondingly, it is a tone often taken by modernist critics, Duchamp, for example, but also Jasper Johns, Warhol, Tom Wolfe (it is often ironical, even sarcastic). To evade these options (cynicism, irony) it suffices to *evade* the paradigm, which can be either obliterated or saturated: both equally are evasive. Obliteration proceeds by destroying the differences in which the paradigm takes root. If the matrix of oppositions differentiates among *underfoot/eye level/overhead,* all will be treated equally. Here the model is Versailles, is Caserta: the floor will be paved with tesserae of semiprecious stones, a parquet of exotic woods will be laid down, Oriental rugs will cover them; frames will be thrown around the doors (these will be constructed of black Mandragone marble), the walls will be plastered with dappled scagiola or covered with frescoes by Veronese, niches will be carved in them (these will be filled with stucco busts); a cove will be run around the ceiling (this will be draped with elastic gilded arabesques); the ceiling will be painted (this will be done by Michelangelo, it will be done by the Carracci, by Tiepolo). Or the model is Katsura, it is the Tearoom of the Shokin-tei, it is a question of bamboo, of oak, of pebbles, of clay, but again, there will be no differences in which to lodge a paradigm such as that operating in our room, there will be no less attention lavished on the mats than on the rafters, no less attention paid the walls than the cabinets. Or, evasion will succeed by *saturating* the paradigm. This is Christopher Idone serving lobster steamed in vodka with a coral mayonnaise on green Fire King (Ingrid collects Fire King): the implicit *high end/low end* is evaded, the distinction is . . . not denied: it is . . . *absorbed*. It is this tone of voice—saturated evasion—that dominates the room. For instance, if there is a paradigm *folk/popular/elite,* the room will attempt to fill all three terms, to evade the paradigm by consuming it (*the Lacandon drum* [39]/*the wicker rocker* [35]/*the Barcelona table* [65]). When a paradigm cannot be saturated in the room, further portions of the home will be encompassed. For example, to saturate the paradigm *folk/popular/elite,* we must draw objects from both sides of another paradigm, *furniture* (35, 65)/*works of art* (39); whereas had we but stepped around the corner into the dining room, we would have encountered a nineteenth-century North Carolina milk chest capable of filling the empty term in the series ———/*wicker rocker/Barcelona table.* The poster on the easel, the wire on the mantel each constitute terms in saturated paradigms.

Obliteration and saturation—though equally evasive of cynicism and irony—take evidently different stands vis-à-vis the existential situation caught in the paradigm. The obliterative tone *denies* differences others observe: thus, "let them eat cake" (though we may as well admit that obliteration is not an elite prerogative: the rooms of those eccentrics in which every cubic inch is crammed with paper bags stuffed with trash equally obliterate locative distinctions, as did the "psychedelic" rooms of adolescents in the late 1960s, in which all surfaces without exception were papered with rock posters or covered with black and purple paint). Saturation *acknowledges* the differences (for example, that ours is a class-stratified society) but denies that *genuine* values (that is, those important to the saturator) are homologously distributed in other classes of objects (thus folk, popular, elite furniture is indiscriminately useful, attractive). Despite acknowledgment, the paradigm is thus fatally compromised (it is occluded), but without denial or being forced into cynicism or irony (the unpleasant extreme here is: innocuousness). It is a purely intensive tone: *no, really!*

None of the voices spoken in the room—the Voice of Comfort, the Voice of High Culture, the Voice of a Certain Easy Formality—is heard as a "pure sound," a single tone. If the principal tone of these voices is normal (essential paradigms are lived out in the room), it is certain to be complicated by off-tones (the paradigm is exploited, it is assaulted, it is evaded). If these are thought of as harmonics, each room—with its unique ensemble—will be heard as having its own distinctive tonal color, a characteristic *timbre*. In our room it cannot be denied that the fundamental is that normal tone of voice *that is no more than expected* (the couch, coffee table, and chairs form a group before the fireplace), but *the* prominent harmonic is that evasive tone of the saturated paradigm (a wicker chair confronts a glass and steel table). Less audible is the assaultive tone of bells deriding the door, the door laughing at the room. The exploitative cannot be heard (it is nonetheless not absent). The complications enrich the tone, the lack of prominence given the assaultive and exploitative make it mellow. The room is a clarinet.

Although this is largely metaphor, at the level where the bells perform as signals it happens also literally to be true: the *clitter-clatter* of the camel bells does mock the well-rounded sonorities of the electric chimes. But in general, because of the consumption of paradigms through saturation, the tone is not that of mockery, but intensification. It is not very much *a sure, you bet* room, but a *no, really!* one.

The room is very sincere.

The Lock

It was George who had tried to visit Herman.

He'd walked up the steps to the house on the corner of Cutler and Cabarrus: "Just come on in," Herman had said, "up the stairs, first room on the right." With his right hand he'd opened the screen door (1), while with his left he'd reached for the exterior knob of the entrance lock. This is a brass-looking ellipsoid two inches across and an inch and a half thick punctured by a hole filled with the plug of the lock cylinder. This in turn is penetrated by the keyhole. It is the plug that turns if the key fits, the key forcing the pins in the keyway to raise the drivers up and into their slots in the barrel which would otherwise—pressed into the plug by their phosphor bronze springs— keep the plug from turning.

The neck of the knob disappears into the door (2) through the exterior rose. There is no escutcheon plate: the lock is a cheap builder's bore-in. With his left hand George turned the knob to the left. When the door stands unlocked, such an action rotates the knob *and* the barrel of the cylinder, both of which are fixed to the door. The knob is connected to a stem whose semicircular end is linked to a retractor capable of withdrawing the beveled latch bolt from its seat within the strike plate. The lock stile of the door is now free of the jamb, and George pushed it open.

As soon as the door escaped from the doorframe, George relaxed his grip on the exterior knob. This permitted the spring in the action to once again project the latch bolt (and the plunger that locks it in place), so that when he turned to close the door

with his right hand—grasping the *interior* knob (identical to that of the exterior except for the replacement of the keyhole with a thumb turn)—it was ready to scrape across the tongue of the strike plate and snap out to hold the door in place once again.

Before this had quite happened, the bells (6) hanging from the neck of the interior knob whacked against the lock stile, producing their characteristic *rattlety-bang,* their *clitter-clatter.*

Then, from somewhere in the house there came a high hulloo and, "Martie! Come have dinner!"

rules   39. "Don't lock it" (Randall, ktAPP, kCON).
            40. "Don't lock your friends out, or your brother" (Chandler, kCON).
            41. "When you close the door, turn the handle . . ." (Denis, ktAPP, tCON, tPRO).

This is how sincere the room is: it is not even locked. Anybody—even a complete stranger—can just walk in.

No, really!

Here is its ideology: that things don't need to be locked up, that people are not thieves. Or, if they are, this is because property itself is theft and therefore indefensible.

This ideology has a praxeology: the door is not locked. More crucially, its locking is proscribed. We have assigned RULE 39 to the control code out of deference to its ability to cover RULE 40 *(don't lock your friends out, or your brother)*. This is a copy (the original? the relevant history is lost) of RULE 9 *(don't lock the screen against family members),* itself the inversion of RULE 6 *(don't open it to strangers)*. This was the specimen for our articulation of the kid-oriented control rule. RULE 39, however, exceeds the urgent specificity of these control rules (which is responsible for their pragmatic utility: *open the door for fresh air*). Its universality pointedly rejects the possibility for control latent in the lock, says: *don't use it.* From perspective of the control code, RULE 39 is uselessly overdetermined.

This is precisely the mark of the appearance code: it exceeds utility, it is involved in something useless, it is ridiculous. What is the point of having a lock with a beveled spring bolt (for instance) if you have to turn the handle every time you close the door (as required by RULE 41)? Here abhorrence of the animal abandon associated with door slamming (RULE 3) can slip within the capacious reasonableness of the control and protection codes. Because the lock works less than perfectly, the latch and plunger often fail to scrape across the tongue of the strike plate: they catch, the door bangs, it shudders, the bells bray and . . . *the door swings open.* Then either there is an exasperated, "Will somebody *please* shut the door?" or the *bang! shudder, bray, squeakkk* . . . is repeated. If it is: "WHEN YOU SHUT THE DOOR, TURN THE (#@!!) HANDLE!!"

What is the problem here? It is that the door may not close, that someone may pass through, pulling it behind him and be up the stairs or off the porch before the door swings back open (violating RULE 22), that having "closed the door" one comes home to find it wide open—admitting summer heat (violating RULE 20) or winter cold (violating RULE 21), that someone sitting on the couch has to get up to close the door after the one who hasn't. It is that the slamming, banging door gradually rips the strike plate off the jamb (2), and hence "this is not the first strike plate to be emplaced and the jamb is a shambles of gouged wood and old hardware." Yet were there no need for these control and protection functions, RULE 41 would nonetheless have been emitted to interdict the animal abandon of the slamming. It is not that the protection and control functions are alibis—they in fact have independent claims to exist—but that they are alibis *nonetheless,* for it is the never articulated abhorrence of abandon that fuels the exasperation animating the rule.

RULE 39 also has its alibis: that no thief would want anything in the house, that if he did he would get in anyhow, that it's a nuisance to have to carry keys. It is true that there is no television in the house, that the stereo system is a quasi-built-in component system scattered over several rooms, that there is no silver chest or jewelry, no golf clubs or guns, that there are not the usual things thieves look for. Yet the thief that *has* entered the house (twice)—he just walked in the wide open kitchen door in the middle of the afternoon—took Chandler's collection of Transformers, took . . . *toys.* Furthermore, all the Woods carry keys anyhow, at least those to their bicycle locks and Denis and Ingrid also carry keys to their places of work. Nevertheless, in that these alibis recuperate the rule in the face of an often expressed incredulity *(what? you don't lock your door?)* they are frequently enounced.

Arrant nonsense for an indefensible ideology. No one believes the alibis (they permit conversation to continue, they keep the Woods from the booby hatch), but sooner accept them than the beliefs they mask. Thieves do exist, people in Raleigh are robbed all the time, their houses are broken into, their homes are invaded. Not only have Chandler's toys been stolen, but so have two hammocks from the porch. Furthermore, the notion that property was theft was articulated with respect to land and capital expressly (not possessions); and in any event the legitimacy of dragging in Proudhon to gloss a quixotic rule is thoroughly suspect. And yet . . . consider the windows in the room (4, 5, 17, 47, 50). Overlooking front and side porches, the street, and a neighbor's dining room, none is curtained. The view out in the daytime is as unobstructed as the view in at night. Compare this to the situation when Denis and Ingrid first looked at the house. On the windows overlooking street and neighbor were Venetian blinds. These were covered by sheers. Over these were drapes. Valences hid blind heads, sheer rods, and drapery headings. One window had its bottom stuffed with an air conditioner. The room was a cave (it was a tomb). Now it's an eyrie. In the matrices of oppositions and affinities mobilized here, *dark/light* is unavoidable: on entering the room for the first time, the light in it is one of the things people comment on. But in the context of *curtained/uncurtained, dark/light* can be no more than a consequent of *closed/open.* In the room these inevitably emerge under a single aegis. This is: *hidden/revealed,* where what is revealed is not only the "*room*" (to the gaze through the window), but the paradigm *hidden/revealed* itself. The *room* declares its openness (it speaks in the performative). RULE 39 enables the door to participate fully in this thematic (the affinitive terms are *uncurtained-light-open-revealed-unlocked*), which moreover is one we have already encountered. It is the theme of hospitality. Denis and Ingrid *are* Philemon and Baucis, their home *is* a temple (etymology: the temple is an open place for augury, it is *unlocked*). Where RULE 6 insisted the door not be opened to strangers, but RULE 19 insisted it be opened to see who it is, RULE 39 prohibits locking the door against anyone. The precarious suspension of a predisposition toward hospitality in a world shrill with the conviction that strangers are never more than potential enemies is tilted here decisively toward hospitality.

The windows are uncurtained, the door is unlocked: *come on in!*

This tilt toward hospitality is a tilt toward a view of strangers, not as gods (the door is not unlocked to escape the flood of divine wrath), but potential friends. This in turn is predicated on a view of man inherently good: *people are not thieves.* Yet it is undeniable that they are (proof by Transformers, by hammocks). These simultaneous avowals (made by the Woods in their unlocking of the front door but nightly locking of that in the kitchen), rooted in different worlds (a world of utopian amicability, a world of resigned hostility) cannot be reconciled, or can be reconciled only through the mediation of a third term. Here it is that people *are* thieves, but that this is no more than a bourgeois alias for the people's right to recover what was always theirs (it is property that is theft). Why in this room should this term be privileged? Consider the paintings in it (41, 64). That over the mantle is of Genovevo de la O, a Zapatista general determined to liquidate the hereditary nobility's monopoly of the land in the Mexican state of Morelos. That on the south wall is of Errico Malatesta, the Italian anarchist who held property the essential means—other than physical violence—by which men are oppressed. In the field generated by their conjunction, either people naturally possess the right to claim that innately theirs (they are not thieves: locking the door would be to perpetuate a calumny), or, they will inevitably become thieves in consequence of the distortions induced by oppression (they are thieves: locking the door would be to identify with the oppressors). That neither Malatesta nor de la O (to say nothing of Proudhon whose "property is theft" most succinctly captures the position) understood by property the table and chairs, couch and paintings of this living room (property was land, it was capital, it was the means of production) is beside the point. The gesture is symbolic (it also says "We will not live in fear"). The refusal to lock the door is nothing other than a behavioral analogue of a certain way of seeing the paintings (the affinitive terms linking windows, door, and paintings are *uncurtained-light-open-revealed-unlocked–social justice*). This significance will be unfolded for the child not through exegesis of the painting but the rule about the door: *don't lock it.* Faith in man will be acquired through practice. The child will consume anarchist philosophy in action.

The site of this action?

The end of the arm where his hand encounters the thumb turn on the interior knob of the lock.

The Floor

In the summertime and the rugs are up, the bright blond floor rolls away from the door like a field of wheat, streaked by the grain as grain by the wind-driven shadows, around the table and chairs, from room to room, off, off into the distant kitchen. A walnut in the corners, it bleaches in the stripes of sunlight to a flaxen refulgence dappled by the dance of monstera leaves moving to the rhythm of the fan.

A cool brightness of oak, sanded, polyurethaned, and waxed. Three hundred thirty-two square feet of it, laid as 385 tongue-and-groove boards (of various lengths) two and a quarter inches wide. Stood end to end these would reach from the ground to just about the top of the CN Tower in Toronto, at 1,815 feet the world's tallest self-supporting structure. Sixty-two feet of shoe molding (in fourteen pieces) bridge the gap to the walls. The classic simplicity of this floor is belied by its baroque contrivance. Give or take a stick or two, there are a further 244 boards and braces, joints and girders implicated in its construction. The floor sits on a subfloor—the two of them are an inch and a half thick—of three-inch oak boards laid across 2″ x 10″ oak joists on eighteen-inch centers. Braced by diagonal oak bridging, the joists span 4″ x 4″ oak girders that rest on brick foundations. These, except where the Woods have had them reconstructed on poured footings, rest directly on the earth, a fine-grained clastic sediment derived from quartzofeldspathic and micaceous metamorphic and igneous rocks (gneiss, schist, granite). There is no cellar beneath the living room, just this chthonian substance, this ramiform endoskeleton of oak, and the air of the crawl space.

42. "Don't stamp on the floor" (Chandler, kAPP, kCON).
43. "Don't use it as a skating rink" (Randall, kAPP, ktPRO).
44. "Don't scratch up the floor" (Denis, tAPP, ktPRO).
45. "Don't mark the floor" (Denis, tAPP).
46. "Please don't spill on the floor" (Denis, tAPP).
47. "Pick up things you leave on the floor" (Ingrid, ktAPP, ktCON).
48. "Don't leave stuff on it so people can trip, like marbles" (Chandler, tCON, kPRO).
49. "There is no leaving of shoes around in this house" (Denis, kAPP, tCON).
50. "Make sure your feet are clean before you come in" (Denis, ktAPP).
51. "Don't walk on it with muddy feet" (Randall, ktAPP).

What is the floor that one should not walk on it with muddy feet?

It is not mud.

That is, it is not a wet slimy debris of water and fine-grained clastic sediments derived from quartzofeldspathic and micaceous metamorphic and igneous rocks, it is not water mixed with earth, it is no part of earth—it is no part of nature—at all. This is what it means to be in a home.

To annul the distinction between earth and floor is to annul the distinction between home and environment, between culture and nature, is to regress, is to return to the clay whence we came. To track the floor with mud is to betray the ground on which human beings stake their difference, is to become animal, to be barbarous (etymology: to be incapable of speech, *like an animal*). What is the floor that one should not walk on it with muddy feet? It is clean. It is next to god.

The floor will constitute itself an anti-mud. Where mud is wet, the floor will be dry *(please don't spill on the floor);* where mud is viscous, the floor will be firm (it will be something you can stand on); where mud is sticky, the floor will be nonadherent (it will be repulsive, it will be slippery, it will be ice); where mud is matte, the floor will be shiny (it will have luster, it will have highlights, it will be glass); where mud is dirty, the floor will be clean (it will be clean enough to eat on [it will be china {it will be glass}]).

The floor will be glass.

What is glass? *It is cooked earth.* To make a floor, we cook the earth. This is metaphor, but were the floor tile it would be literally true. In our room, oak has been transmuted into glass through . . . finishing. This is a labor of culture. The oak floor—already stripped of every sign of nature but the trace of growth preserved in the grain (trees have been killed, milled, reassembled as flooring)—is sanded, is varnished, is waxed. This vitreous surface is continuously attacked: dust builds up on it, dirt, mud, leaf litter is tracked across it, it is scratched, it is marked, the legs of the couch gouge it, it is abraded by the passage of numberless feet. To retain the sign of the glass, this attack is resisted. Broom, dust mop, long-handled brush to get under the radiators, and three vacuum cleaners (a mini, a "broom," a rolling canister) constitute a very armament of defense. Denis engages the enemy with the mini and the "broom" on a daily basis. Ingrid dust mops once a week and vacuums—"as little as possible"—twice a month. Uncounted hours are absorbed in the effort. When the pollen drifts yellow on the floor, when the weather is bad (and it is muddy outside), when the kids have friends over and there is much in-ing and out-ing, it can explode into a veritable hysteria of mopping, of zapping with the mini-vac, of swiping, of picking up. Twice a year the offensive is taken. All the furniture that can readily be moved is . . . *out of the room.* On her hands and knees Ingrid attacks the bad spots with paste wax and steel wool. She then waxes the floor with the electric waxer and "Johnson's original blend of hard finish waxes" which form a "lustrous, oil-free dust and water resistant coating. Additional waxing adds rich lustre." She polishes the floor half an hour later. Finally she buffs it with the brushes covered with felt pads. The floor shines. It gleams.

What for?

There are three reasons. The first we have seen. The shining gleaming floor is glass, that is: not mud. It has been precisely: *cooked* (in an oven of mops and brooms, vacuum cleaners and wax). It is culture: *this* is what we stand on. The oppositions here are classical (they are those between female mud wrestlers and Cinderella and her slipper of glass): *nature/culture, raw/cooked, earth/glass, wet/dry, viscous/firm, sticky/nonadherent, matte/shiny, dirty/clean.* Through the affinitive terms *culture-cooked-glass-dry-firm-nonadherent-shiny-clean* the floor joins the doorframe (a glossy enamel white: *keep your hands off it*), the window in the door (plate glass: *don't breathe on it*), and the glass in the sidelights (don't let your friends put their noses on the windows) in an increasingly univocal syntagmatic structure. What we wish to observe here, though, is not the matrix linking the floor through the beveled glass to the front of the house (to the face, the mouth, guests, eating as ritual, social pretension), but the system of affinities that asserts in this larger scheme the hegemony . . . *of the clean.* This is the second reason that the floor must shine, must gleam, not that it be clean so much as that it be a surface capable of taking *the mark of the dirty* (so that it can participate in the economy of the clean). Nothing is more *natural* than the *dirty* (before food is *cooked,* it must be washed). Nothing is more capable of signifying the cultural than the absence of dirt, an absence the more readily paraded on a gleaming, reflective surface (polished silver, a crystal wine glass, the finish on a fine automobile). Here is lodged the opposition *carpeted/wooden,* where what distinguishes the carpet is its propensity (some would say its ability) to hide dirt. Relevant here is Ingrid's mother's refusal to countenance wall-to-wall carpet because it was filthy. If carpet you had to have, enough space had to be left around the edges to run a dust mop; but the institution was inherently one of *hidden dirt* (as carpet beating amply demonstrated). Ingrid has an equivalent abhorrence of linoleum: *who knows what's happening to the floor beneath it* (theme of the rotting, another route to nature). Here, where *carpeted-linoleum/wooden* recapitulates the paradigm *hidden/revealed,* the floor is seen to participate in the thematics of openness focused on the lock, but now it is a question of vigilance: *what can't be seen is probably dirty.* Therefore, the floor *exposes* its cleanliness. Much is at stake: in the shallow space between the dirt below and the wax above is inscribed no less than the rise of civilization. Though culture is most present *on the surface* (in the lack of dirt on the mirror finish), the floor itself is also of culture. Thus, when it is stripped, it does not return to nature (though its declension *is* changed). With the subfloor, however, culture is all but forsaken. The subfloor lies . . . *underneath,* participates already in the chthonic. Nothing but the air of the crawl space separates it from the dirt, a dirt, let us insist upon it, that—simultaneously signified and signifier—is not merely the not-clean but the inevitable end-state of all cultural degradations. From the waxed to dirt: everything that is not culture tugs in this direction.

But culture is a ceaseless resistance. Whatever is pulled toward dirt must be cleaned—if only to be dirtied again. Because culture is a resistance not a conquest, cleanliness is never more than the *temporary* absence of dirt. The system is perverse: to be of culture is to erase the mark of the dirty, but to be erased, the mark is required. Paradoxically, the work of culture will be to circulate dirt. In the home, four major circuits can be identified (there are innumerable minor ones). Dirt is removed from—but ceaselessly reclaims—bodies, clothes, dishes, house. This means that in the life of the child, the floor and his cup, his socks and his face, can with equal effect comprise sites for the cleaning that is his culture. Because nature is implacable, culture is not obliged to *impose* the mark of the dirty. This happens, as we say, quite naturally. In fact, it is all culture can do to maintain itself. Insofar as the child is barbarous, he is natural; insofar as he is natural, he is an agent of the dirtying. In the economy of the clean, the rules subserve dual functions. Insofar as they are followed, they mitigate against the dirty (the rules clean); simultaneously they instill in the child a commitment to the clean (the rules instruct): as they stay the hand of nature, they teach the ways of culture. Thus, *don't walk on the floor with muddy feet* keeps the mud off the floor (it is concerned with the appearance of the thing, it is coded tAPP) at the same time that it attempts to inform the child that a human being is that which does not track mud in (it is concerned with the definition of the child, with the way he appears, it is coded kAPP). Because *make sure your feet are clean before you come in* no more than acknowledges that there is more to nature than mud, it is similarly coded (ktAPP). Because *pick up things you leave on the floor* no more than acknowledges that kids have hands as well as feet, it too is lodged in both sides of the appearance code. But because it further acknowledges that these things may have roles beyond that of "muddying" the floor, it is also lodged in both sides of the control code. This assignment reflects the reality that in order for us to *use* things, they need to be *put back in their regular place* (kCON—it is the kid who needs to put them back); but at the same time it pays obeisance to the reality that out-of-place things can be dangerous (tCON). Thus Chandler notes *we don't leave stuff on it so people can trip, like marbles* (tCON), where the additional assignment to the protection code acknowledges that the kids are themselves the ones most likely to trip (kPRO). On the other hand (in response to the repeated *has anyone seen my shoes?*), in *there will be no leaving of shoes around in this house,* Denis emphasizes the need for *shoes* to be put back in their regular place (tCON)—where for Denis this behavior is also (as with the slamming of the door) the mark of an intolerable behavior (hence kAPP). The remaining rules relate either to the appearance of the floor—*please don't spill on the floor, don't mark the floor,* and *don't scratch up the floor* (where the secondary assignment to the protection code refers to the consequent splinters, as injurious to floor as child)—or the child. Though neither Denis nor Ingrid recognize either *don't stamp on the floor* or *don't use it as a skating rink,* certainly both could have been enounced. When the floor has just been waxed, it *is* slippery; and the protection coding of the skating rule reflects the knowledge that the sliding kid will either fall, smash into something, or both. Before mats were put under the rugs, guests were insistently told *be careful walking on the rugs, they're very slippery,* to prevent their sailing awkwardly about the room.

There will be no treatment here of the rugs—they were not on the floor when we took our inventory of the room—but they are the third reason the floor must be shiny, must glean: to be a suitable surface against which to view them, the Navajo, the Afghan kelim, the Irani boukara. For Denis, this is how it is, what one does: one lays fine rugs on highly polished hardwood floors. It is just the way people live. This is an historic endowment, an instillation of parents whose homes had to have, *a minimo,* walls big enough for the big paintings and hardwood floors for the rugs. They were renting space for an art gallery in which, incidentally, to live. Here is a crux: *showcased rugs/visible dirt* recapitulates the paradigm *Denis's side of the family/Ingrid's side of the family* but the floor required by each runs from one side to the other through the virgule, just as in the home the floor runs without break from the living room into the dining room, uniting, on the plane of the feet at any event, the pleasures of the table with those of the mind through the agency of . . . *the clean.*

The Radiator by the Stair

When the door stands open in the spring, the radiator by the stair is almost invisible. Swamped by the gush of light at the window above it (17), and that abounding from door (3, 4) and sidelights (5) beside it, it is little more than a shade or the tone of a shade in the space behind the door (2). But even when the door is closed, the radiator is hard to see. Not that it is not there. Standing two inches out from the wall, three and a half feet long, two feet high, and six inches thick, perfectly centered below the window in the space at the foot of the stairs, it does everything it can to draw attention to itself but whistle.

Which as a radiator in a forced-circulation hot-water system is one of the things it cannot do. What it can do is heat the stairwell and its corner of the living room. This it does by circulating hot water under pressure through the nest of tubes of which it is constructed. There are 110 in this case—more than in the side radiator (46), fewer than in the front (45)—in vanes or tiers of five, tied together through the thickness of the radiator at the top, bottom, and middle; but along its length at top and bottom only. Each tier is an individual casting, its five tubes separated by air space, but otherwise treated as a single (elegant) entity. The outer two have "legs" (and holes for the bleeder valves) but are otherwise indistinguishable from the twenty they sandwich. The necks where the tiers are connected project so that each tier is as far from the next as it is thick. The whole thing is held together by four tie rods. Driven by a pump (or circulator) at the furnace, water is brought to the radiator through a system of main and branching supply lines. A supply riser (or inlet pipe) brings the water to the shut-off valve at the top of the radiator. This is operated (rarely) by a black Bakelite handle. Transferring its heat to the radiator, the cooling water works its way down to the return riser at the bottom of the other end. A bleeder valve at the top of this leg is operated (rarely) by a key. Some of the heat transferred to the radiator body *is* radiated out into the "room," and some of it—as when you sit on it in the winter time (it is never very hot)—is even transmitted by conduction. But most of it is transmitted by convection, which of course is the point of getting the radiator up off the floor, so that the air can be pulled past its ramified surface, heated, and shot up off the wall, up the stairwell, and out into the rest of the "room."

Originally the radiators were painted a metallic gray. Of those on the ground floor, only that in the side room has not been repainted, most of these in what must have been when new a glossy white, now yellowed to a dirty cream. The paint is badly chipped on the radiator by the stair, showing here metallic gray, there rusted metal. The space between the tiers is shadowed: dirty cream, dark shade, dirty cream, dark shade . . .

rules   52. "You're not to climb on the radiator by yourself" (Denis, kPRO).

53. "We can't stand on it" (Randall, kPRO).

54. "Take items on the radiator upstairs when going up" (Ingrid, ktCON).

Immediately the radiator is revealed in another light. It is not a source of heat, it is a stool, it is a table . . . it is a surface. This reflects above all the fact that as a hot-water radiator it never gets as hot, say, as a steam radiator, which when Denis was growing up melted a box of crayons into an indelible mess, and even more recently (in the last place Denis and Ingrid lived) warped a bunch of records into unplayability, the Ingrid Haebler–Ludwig Hoffman *Mozart Piano Music for Four Hands* (never replaced!), the Landowska *Haydn Sonatas,* Miles Davis's +19.

Anything can be placed on the radiator by the stair with impunity, but what *is* placed here are things that have to go upstairs *the next time anyone has to go up for something else.* Therefore, *when you're going upstairs* (this is the kid-directed side—at stake is the development of a sense of responsibility for the operation of the home), *take the things on the radiator up* (this is the thing-directed side—it will be toilet paper for the bathroom upstairs). Because the radiator is shallow, wide things like the basket of freshly laundered clothes will be put on the floor at the foot of the newel post (10). Because the radiator is farther away, or because it is not a smooth continuous surface, small things—pens, pencils, a book—will be put on the top of the newel post itself (with respect to which a similar rule applies).

Both the other rules are defunct. They were emitted with the intention of keeping the kids from killing themselves when, seduced by the window (17) out of which they could not see, they wanted to climb on the radiator to do so. Given the difference between Randall's and Denis's versions, it is worth speculating on its original form. Ingrid probably said something like, "Guys, let's not climb on the radiator, okay? unless Den or I am around to help you? Okay?"

The Newel Post

The newel post is most itself on a gray day. Then it is an ebullition of edges chiseled sharp by the cool and steady light. Sunlight, scattered by the white enamel paint, bleaches it to a uniform brightness; and lamplight is too diffuse to limn the shadows; none of which in any case is cast by more than the thinnest of fillets, the shallowest of scotias, the least protuberant of toruses. Yet there is an immense number of these, the seventy-three visible pieces of wood—mostly moldings of varying complexity—contriving to articulate 337 individual surfaces, each with an edge, clean in the graylight.

The newel is square in cross section where it rises from its seat among the framing as a plinth vertically elongated to receive the closed string of the stair. Shoe molding smoothes the transition from the floor to the plinth, as bolection molding eases that from the plinth to the paneled shaft of the post proper. Above a second course of molding (ovolos bracketing a torus) stands a shorter section of paneled shaft, adapted to receive the rail—the banister—of the stair. This is surmounted by a cornice (cavetto, double torus, and ovolo) capped in turn by a beveled block raised by a scotia from the thinnest of plinths. For all this, the effect is quite simple, Tuscan, forthright, and square, without being either plain or simple-minded.

It has about it something of the air of Giotto's campanile for the cathedral in Florence, especially when the day is gray and a Kleenex box isn't waiting on top for a lift upstairs.

rules 55. "Don't sit on it" (Chandler, ktPRO).
56. "Don't stand on it" (Randall, ktPRO).
57. "Don't bang into the newel post" (Denis, tPRO).
58. "Do you have to swing on the newel post like that?" (Denis, tPRO).
59. "Take up things on the newel post when you're going up" (Ingrid, ktcON).
60. "Don't leave anything on it" (Randall, tcON).

The room is a competition. Enter it as you will, your eyes will not fail to rest on the doorframe (on its 63 square feet of bright and lively wall), on the floor (on its 332 square feet of bright blond oak), on the stairs (on the 52 square feet of its bright white balustrade).

Why do we not see the radiator?

Because our attention is absorbed, consumed by the doorframe, by the floor, by the stair supported, announced, and summed up in the newel post. Where the radiator squats against the wall, the newel post will tower in the middle of the action. Where the radiator is a dull chipped tan, the newel post will be a smooth and shiny white. Where the radiator hides behind the door, the newel post will bristle before it. The radiator is an unnecessary evil (*oh, those old radiators* [that is, not only ugly but *no air conditioning*]); the newel post a decorative boon (*I just love these old houses!*). The one reminds us of our body (of nature); the other (generously) alludes to our mind (to culture).

None of this is apparent in the rules. If Ingrid wants people to *take items on the radiator upstairs when going up,* she is no less concerned that they *take up things on the newel post when going upstairs* (no countermand, Randall's version is a succinct negative of Ingrid's comprehensive positive). If once the kids were not to climb or stand on the radiator, they remain forbidden to sit or stand on the newel post. The injunctions against banging into (which in any event Chandler didn't understand)

and swinging on (which Randall refused to admit was enjoined) no more than reflect the reality that, wooden (instead of cast iron) and in the middle of things (instead of off to the side), the newel post stands in greater peril than the radiator from barbarian assault—especially when it is acknowledged that to use the stair is to grab, stop yourself with, swing around, or otherwise all but rip the newel post from the floor. But this difference is trivial: from the point of view of the rules, neither will be a jungle gym, both will be tables.

Where then are the missing rules, the ones to mark the alliance of the newel post with the doorframe, sidelights, window, and floor? The rules that protect the white, the bright, the beveled, the shiny, the excessive? The rules enjoining touching (RULES 24, 26, 31), smudging (RULE 27), scratching (RULE 44), marking (RULE 45), spitting (RULE 35), and pounding (RULE 34)? Here, at this break point in the traffic of the room, at this annunciation of the origin of the ascent to the second floor (or the termination of the descent), at this capital letter inaugurating a novel syntagmatic unit (or exclamation point bringing it to a conclusion) . . . they are absent. This evaporation of the appearance code marks a concession to the pragmatic which above all else dominates the meaning of the stair. The stair will be a thing which *works* (the woodwork postures, it *appears*). Hence, here, no "Please Don't Touch."

Please touch.

Don't fall.

The Banister

Late on a day in June and the sun is aflame on the floor, the underside of the banister takes on the buttered glow of the chin of a kid being asked if he likes butter by another who holds a dandelion there.

The glow is strongest where the banister meets the newel post, fading through cream and milk to the blue tones of skim where the banister slips past the ceiling in its flight to the second floor. It never gets there, stubbed off at the newel post on the landing, twelve feet three inches from where it began. The space between the banister and the closed string of the staircase is filled with balusters. These are twenty-seven in number, not quite two feet from top to bottom and square in cross section like the newel post (but only one and a quarter inches on a side). On five-inch centers, they strike one as strong, but not clumsy; solid, but airy. Taking the newel post into account and the closed string, it amounts to fifty-two square feet of bright white balustrade.

The banister serves two gods, the hand, for which it is convenient to be curved, and the balusters, for which it is convenient to be flat. Two of the three and a half inches the

banister is deep are devoted to reconciling these divergent formalisms, wedding an oval to a square through the grace of molding.

rules   61. "Don't play around the banister" (Denis, kPRO).
                62. "Don't slide down it" (Chandler, ktPRO).

The stair will work.

It will not be like the screen door which, in its attempt to control access *(don't open it to strangers),* could no more than play at working.

It will be like the door, so encrusted with rules that to no more than brush by the stair will suffice to knock a couple off *(do you have to swing on the newel post like that?),* though here it is the enticement of the banister that is denied, the twelve-foot slide ("I got in trouble once," Randall told Bob), the partial screen, the balusters to clack as you run up and down the stairs (NO RUNNING ON THE STAIRS!!), the palisade behind which lurks—*don't play around the banister. How many times do we have to tell you that?*

The stair will work . . . by not playing.

XII · THE LEVEL OF THE CODES

To assign a rule to a code is not to answer the question, "What purpose does this rule serve?" but to forgo asking this question at a certain level. This level is that of the code.

A rule is always in favor of something. *Don't push things through the holes in the screen* is in favor of the screen. It elevates the integrity of the screen over the desire of the child to noodle at the unraveling fiberglass: it favors the screen. To ask what purpose this serves is to ask, "This favoring of the screen: what does it favor?" This cannot be answered at the level of the rule. As of any organization, it is possible to say, of any level, that its *mechanism* lies at the level below (in this instance, the anatomy of the kid, of the fiberglass), but its *purpose* at the level above (in this instance, that of the code). We will say, for now, that the *protection of things* (tPRO) is what this rule favors. There is, of course, no reason to stop at this level. For example, one could ask, "This *protection of things:* what does *it* favor?" There is no way to answer this question at the level of the code, whose mechanism (the rules) lies on the level below, but whose purpose must be sought at the level above. We have denominated this the level of the Voice. We will say, for now, that a Well-Ordered Life is what the protection code favors. "And this Well-Ordered Life: what does *it* favor?" The levels ascend (the next is that of the West, beyond which lies human culture, sociobiology, thermodynamics, physics, Nature). To assign a rule to a code is to terminate this ascent *at a certain level.*

What is this level?

It is that of the family, it is that of the living that is the home, it is that of the room. Below this level are individuals and individual things: this is the level of the rule. Above this level is social class: this is the level of the Voice. The relays that enable *class* values to permeate *individual* action (for the Voices are nothing but constellations of values seen from a certain perspective), that permit the Voice of Class to be spoken in the *rules of the home* are the *codes of the family*. They do this by translating the abstractions of class *(you simply have got to start having better manners)* into the concrete actions of people in a world of things *(we do not eat with our elbows on the table)*. They are in this way not unlike the assemblers, compilers, and interpreters which translate computer instructions in compiler or assembly languages (at the level of class: utterances of the Voice) into machine language (rules) capable of being acted upon by the (magnetic) switches comprising the central processing unit (kids). It is as silly to expect a child to understand the dictates of class as it is to expect a computer to act on English: each is ultimately obligated to be transformed into a stimulus susceptible of reception at the level of the intended action. (There are, of course, as many levels below that of the central processing unit and the kid—silicon chips, neurons, molecules, atoms, elementary particles, Nature—as there are above that of class; but to assign a string of words to the category *rule* is to terminate the *descent* at a certain level too.)

The limits of this analogy are those established by our use of the word *code*. Although the code translates the Voice into rules precisely as a compiler translates compiler language into machine language, it nonetheless lacks the rigorous specificity of that compiler. Not only does it fail to meet the requirements of a mathematical "group" (as a code such as a cipher does), but, as this may imply, it dispenses with the code book too. The resultant play in what are no more than associative fields to begin with permits the Voice of Class (and of course everything above it) to speak in the rules of the home without ever being heard as such: individual behaviors seem to be required, not by the Voice that demands them, but by the codes that relay them. It is this mask of the codes that above all else naturalizes the culture of the room. Why *this* rule? And one points to a code. From the other side, this has the effect of naturalizing the precepts of class: by obscuring the rules which instill class behavior, the relay of the codes allows class behavior to appear . . . unconstructed, natural, that is . . . genetic: *breeding shows*.

To assign a rule to a code, then, is to justify the rule at the level of the family, at the level of the living that is the home, at the level of the room. Since this justification has much to do, it will rarely be simple. Instead, it will invoke other rules and codes in a ramifying network whose circularity will succeed in sealing off inquiry at the level of the family. Relentless pressure on this essentially self-referential system will elicit a *because that's the way we do it* or *it's only common sense* but rarely will the seal be broken that insulates the level of the family from that of social class. Why, for instance, *don't [we] push things through the holes in the screen?*

The answer is so obvious that any but the most impertinent (or bored) child will see it immediately: *because we don't want to destroy the screen.*

But why not?

This question can be handled many ways. One is to insist that the answer is known, or could be known if only the child would make recourse to that fund of shared common sense: *don't be a smart aleck!*

No, really, I'm serious. Why not?

Again a choice presents itself. One answer is *because it looks tacky.* Another is *because then the bugs will get in.* Both are connected to other rules (the justification may have originated with them), the first with *don't push on the screen* (why not?— *because it looks tacky,* though through the intermediary of *because it will come out and then the bugs will come in,* this is additionally connected to *don't push things through the holes in the screen*); the second with *close it every time you go through* (why?—*so the bugs won't come in*).

What's wrong with having bugs in the house?

They'll bite you.

And so on and so on, round and round the circle of the codes. In this scenario— simultaneously hypothetical and inevitable—the parent first exploits the protection code (tPRO), switches to the appearance code (ktAPP: a kid noodling unraveling fiberglass is no less offensive than a saggy holey screen), runs to the control code (tCON: keeping the bugs out *is* what the screen in the door is for), before concluding with . . . the protection code (kPRO). So doing, the parent has shifted objects (from *t*PRO to *k*PRO), hereby revealing to the child its own putative interest in protecting the integrity of the screen. This appeal to enlightened self-interest is not, however, the point. What is at stake is not a rationalization (which in any event could only be followed by an older child), but a demonstration, and this not in the logical but the political sense: *reasons on the march!* A fog of argumentation is allowed— encouraged—to obscure the inherent circularity of the argument, one apparent consequent follows another, and if the chain is long enough who would be sufficiently intrepid to test for weak links? The child emerges from the encounter dazed by the spectacle of a rationality that comes gradually to be taken for the frame of the family, a frame within which all further justifications will be found to have an a priori plausibility: *it's just like the screen door* . . . And the child nods his acquiescence.

And doubtless it *is* just like the screen door, as doubtless bugs *will* come through the holes in the screen. The system's effectiveness derives from its truthfulness: the noodling kid *will be* eaten alive. What passes unnoticed is the failure of this truth to exhaust the achievement of the rule, which not only *protects* the family from bugs but *projects* for the world, through the taut and seamless screen, the façade of a Well-Ordered Life (everything neat and tidy, the kids under control), of a certain Easy (the welcome of an open front door) Formality (the screen door offsetting any implication of wantonness latent in an unguarded entrance), of a kind of Comfort (fresh air). It is not that these Voices are unheard (they are what one listens for), but that they are unheard *in the rules*, that they seem to exist independently, effortlessly, as the effects of grace achieved by Fred Astaire seem to exist on the screen, his by rights, because of who he is (a graceful person), not as the contrivance, the artifice, the result of practice, sweat, and choreography. So the Well-Ordered Life, a certain Easy Formality, and a kind of Comfort have also to be seen, ours by right, because of *who we are* (that kind of people).

It is not something we have done. It is fate. It is manifest destiny.

XIII · THE CONTRIVANCE

This spectacle of rationality has an ally, a counterpart, a partner. This is the spectacle of functionality. In a phrase, *the home works.*

Day in and day out. People get up and go off to school. Dinner is put on the table. Switches control electricity. Knobs control gas. Day after day people fail to crash through the floor into the basement. How does this happen?

Effortlessly.

It happens without effort, without apparent difficulty, without strain, it happens . . . naturally. Everything contrives at this illusion of smoothness (nothing jerks, *the home is well lubricated*) which, in its all-embracing, unrelenting, self-illustrative, unendingly reiterative way constitutes in itself a profound (if silent) argument for the ways of the home (its smoothness of operation will be sufficient justification for its being, like a law in physics). Does this mean there is no strain? On the contrary. It means the strain has been banished to the basement (or the kitchen or the privacy of the bedroom . . . *or the environment at large*), hidden beneath coats of paint, buried in the walls: it is off-screen in any case (it is below the floorboards).

Exactly: it is below the floorboards. Randall—gangling preteen—hurls himself downstairs CLUNK CLUNK CLUNK CLUNK CLUNK: *THUMP!* The floor at the foot of the stairs does not give way, Randall does not crash—screaming—into the abyss below, once again . . . *it works!* And everything is like this. Denis grabs the newel post on his way upstairs: *it does not pull out of its place in the floor.* Ingrid opens the door: *it does not fall off its hinges.* Winter comes: *and the family does not freeze to death inside.* The amazing thing is—*nobody notices.*

This is the contrivance.

But the strain is there—literally, in the engineering sense—whenever Randall thumps onto the floor at the foot of the stairs, every time he walks on the floor, every time any of us walks on the floor: in the floor vibrating like a string, in the compression waves washing through the floor to the joists and the girders and through these to the foundations, in the bricks shrinking from the blow, in the earth receiving, finally, through the house, the burden of Randall's weight. The strain of the gracious life, which the codes try to pretend is *not* the manifestation of a position in a class-stratified society, is absorbed by a heavily greased contraption whose crabbed elaborateness—once observed—is vertiginous. It is easy to lose count, but there are no less than 244 pieces of wood supporting the floor, and more than 800 other pipes, tubes, springs, splines, boards, handles, bells, roses, locks, latches, and pieces of glass—to stay on the level of the family handyman, to accept a certain degree of preconstruction, to overlook the nails and the bricks—in no more than the fraction of the room we have surveyed. The effect of effortlessness in the operation of the home is achieved by displacing effort into this wealth of screens and screws, moldings and mullions, rails and radiators, into this cornucopia of paraphernalia or *through it* to the appropriated labor of others, the energy resources of fossil fuels, or both. The effect of effortlessness is thus no less an artifact of class than the system of justifications circulating through the codes, from any evidence of which it is no less carefully insulated. Hence this effect of effortlessness will not be attributed to appropriated labor or appropriated oil (secured at the cost of war), but to the skillful management of the householders, to their personal ambitions and abilities. In the case of both codes and contrivance, the enabling structure is made to disappear—the social superstructure, the physical infrastructure—so that the welcoming gesture of *swinging open* a screen door with *concealed hinges* is seen as no more than a coincidence of personal preferences, not as a coincidence toward which the entire system conspires. But the physical mechanisms permitting the employment of the door as a prop will never be more acknowledged than the social mechanisms mandating the effusive greeting that demands it. Cause will be banished. Effect will dance alone upon a stage of its own devising . . .

	CAPITALISM	classes	VOICES
Levels:	CONTRIVANCE	families	CODES
	EFFORTS (labor)	kids (individuals)	RULES

But this effect is a sham. Precisely as the Voice of Class speaks through the codes, the economy of capitalism is manifested through the contrivance. As we have seen from the very beginning, the *thing* is subject to an entire cacophony of degrading processes reaching conclusion only with the *thing's* demise. Unless it be preserved—in a monstrous effort of repair, mending, picking up, putting back together, sanding, washing, polishing, cleaning, straightening, setting, fixing—and protected—by the force fields of the rules. And even so it is sooner or later gone, demanding, at the very least, replacement, if not the endless supplementation toward which a consumerist society ferociously propels us. Nor is it that this presence of capitalism is unfelt (it is the touchstone), but that it is unfelt *in the effort,* the effects of which are admired precisely in the degree to which they appear *uncaused,* the miraculous gift of a bountiful world, neither appropriated not coerced, ours because of *who we are* (deserving).

They are not things we have labored to produce. They are a boon.

The Stair Treads

We know the stair treads best through our feet. Not that we don't see them, but that even when we *cannot see,* we use them as though we could. In the dark we stick our feet out—and there they are, fragments of floor, dribbling down from the second or floating up from the first, mingling, the one floor turning into the other (but who will say where?), at night, beneath our feet.

Their number is elusive. It is seventeen steps from the first floor to the second, but there are only fifteen treads, the thirteenth step (going up) taking you onto the landing, the seventeenth onto the second floor. The treads are oak boards, thirty-five by eleven by one, with generous nosing, nailed to a pair of stringers that run up beneath them. The closed string is merely decorative. That is, while it is not implicated in the support of the treads (it supports the balustrade), it does conceal their ends from view. There is not even the intimation of a wall string, but where the tread does not quite meet the wainscoting, sections of shoe molding have been emplaced. This molding, which juts awkwardly beyond the nosing, is not found on the three treads above the landing that make the turn to the second floor.

Which, as we say, is a whole 'nother story.

rules   63. "Don't play on the stairs" (Denis, ktcon, kpro).
64. "Don't stomp on them" (Chandler, kapp).

In the room, the distinction between play and work is nowhere as well differentiated as on the stair. The stair will work . . . by not playing. But what is this work? It will be to violate the virgule in the paradigm *downstairs/upstairs,* which, in a two-story home is the physical expression of the paradigm *public/private.*

To enter the room through the door (2) is *to see* at once to the kitchen through the continuous volume of space underlain by the floor (8) (*into* the kitchen actually, all but through it to the yard out back); and *not to see* around the bend in the stairs (15) to the second floor at all. The eye is pulled through the first floor; it is stopped at the landing to the second. It is this difference between *visible* and *invisible* that instantaneously maps *public/private* onto *downstairs/upstairs*. The floor (8), that on the plane of the feet could elide differences between *front* and *back* and *Denis's side of the family* and *Ingrid's side of the family*, will not be able to bridge the oppositions between *public* and *private*, between the *first* and *second* floor. Here the floor will be broken, shattered, fragmented into treads floated up into space: the *public/private* paradigm will break the floor. Paradoxically, these *fragments* will succeed in bridging the two domains: the public and the private will be linked by broken floor (tread carefully).

This linkage will be perilous. The space that it bridges, that between stranger and friend, is—unlike that between formal and informal (elided by the intact oak of the ground floor)—never casually crossed. Here the passage will be trebly marked. First, the movement will be upward along a broken path. Then it will take the ascendant from the wide panorama of the open stair—out over the landscape of the living room—into the narrow defile of the closed stair above the landing, debouching onto a closed hall. This constriction will be metaphoric: socially, a field *has* been narrowed. Finally, as on any mountain climb, there will be the possibility of falling (surrealists advocate the removal of banisters so that users of stairs may learn to overcome vertigo). Intimacy will not reduce this danger, but be confirmed by it, and will increase the frequency with which it is encountered. Family members will face it so often as to come to ignore it; consequently they will be the ones to fall (as most recently Denis, cartwheeling from the space above the landing to that at the bottom of the stairs). Because any misstep can induce this trauma, the possibility of misstep will be reduced. This will be accomplished by prohibiting the use of the stairs except *at need:* something will be *required* from the other part of the house. This will be an object (bed, a book, something from the refrigerator) or a condition (seclusion, company). The stair will never be exploited *for itself*—it will not become a toy, an object of pleasure (running up and down for fun)—or *as other floors are:* the treads will not be surfaces on which one does anything but walk (they will not be played on, nothing will be left on them, things will not be put on them, not even rugs, not even runners).

Even this use will be circumspect: *stairs are for going up and down* (RULE 73). The refusal here to specify the form of *going* derives from the fear that any form could evolve into excess: one will not: stomp, clomp, run, jump, slide, bump, crash, thump. One will just . . . go.

The Risers in the Stair

We know the risers best with our eyes. As we climb the stairs they constitute the wall that recedes with every step. Yet our eyes tell us the risers do not lack a knowledge of our feet, each square foot and a half a billboard inscribed in the language of the toe.

Seventeen risers, painted white, likely pine. Violating the cardinal rule of stair construction, they vary in height from seven to eight inches, an inch of which is lost to the thickness of the tread. A little piece of molding has been tacked beneath the nosing of the three bottom treads, but otherwise the risers fall before the climber unadorned.

Except for the epigraphy of the toes.

rule   65. "Don't kick, mar, the risers" (Denis, TAPP).

If there were any question of the kids standing there kicking the risers in fits of *what's-to-do?*, RULE 65 would be coded *kt*APP, but there isn't. Not the Death Star was as well protected as the stair, enveloped in its field of protective rules (to allude only to these) that makes it all but impossible to even *get* to the risers, much less stand there and kick them (see RULES 55, 56, 57, 58, 61, 62, 63, 66, 68, 69, 70, 71, 72, 73). *Don't kick, mar, the risers* no more than acknowledges the risers as *wall*, subject, as such, to the authority of the appearance code which all but exclusively dominates its meaning (see RULES 74, 75, 76, but especially 77). Yet the powerful pragmatic inhabiting the stair will even here interpose itself: the force of the rule is completely evacuated by nothing less than the toes' writing on the risers. When it comes to the stair, *que sera sera,* as long as it works.

The Landing

About the landing there is something of the cloud in a baroque painting, above the action, yet not quite in heaven.

Except that no one does time here, it's almost purgatorial, a pause, a turning in the way . . . on the way up into the dim, on the way down . . . into the light.

A real room, it has its own floor, walls, ceiling and newel post, but as if to sign its difference, the oak boards run counter to those of the first and second floor, wainscoting covers most of the walls, the ceiling slopes, and the newel post is capped with a globe instead of a beveled block. Respectably enough, a shoe molding smooths the transition from floor to walls.

Now and then of an evening Chandler will sit on the landing like an angel of Caravaggio, unseen, but seeing . . .

66. "Don't jump on the landing" (Denis, ktPRO)
67. "Don't stomp on it because the record will scratch" (Chandler, ktPRO).
68. "No playing on the landing" (Ingrid, tCON, kPRO).
69. "No leaving things on the landing" (Ingrid, tCON, kPRO).

The landing is the site of an amphibology: caught up in the "stair," it is constructed as a "room." Any attraction it might have had for play is doubled by this doubling of its potential interpretations: it is a room in a stair, it is a stair-room, it is an eyrie, it is a crow's nest, it is a conning tower, it is a cockpit, it is the peanut gallery, it is a tread on a stair, it is a slip, it is a disaster, it is the dread of the ambulance, of concussion. It is a twilight place, neither day nor night; it is a werewolf, neither animal nor man. Its ample size and construction insist on the paradigm *stair/landing-floor,* but its minuscule size and location insist on the paradigm *stair-landing/floor.* Never a middle term, it will shuttle from one to the other.

Therefore it will be treated as both. *No leaving things on the landing* is a floor's rule: compare *pick up things you leave on the floor* (RULE 47) and *don't leave stuff on it so people can trip, like marbles* (RULE 48). These take for granted that things have been put there, asks no more than that they not be left behind. This is in keeping with a fundamental difference between adults and kids: adults play *around a table;* kids play *on the floor* (where else? it is closer to the animal, to the dirt). On the other hand, *no playing on the landing* is a stair's rule: compare *don't play around the banister* (RULE 61), *don't play on the stairs* (RULE 63), and *no playing on the stairs* (RULE 70). These take betrayal by the stair for granted, assume that it will trip you up, throw you down, land you at the doctor's. Famous last words will be: *we'll be careful:* it cannot be played on safely: it will not be played on at all. The rules annihilate each other (they assume *play* and *not play*). Yet they coexist.

To understand where we are, it may be useful to recall the position of the dining room. Here the spatial distribution of windows with quarrels in them distinguished the living room together *with* the dining room from the rest of the house; whereas the paradigm *Denis's side of the family/Ingrid's side of the family* made the cut between the living room and the rest of the house. If we restrict our interest to the three rooms strung along the axis stretching from the front door to the back, we can notate this in the shifting opposition *living room–dining room/kitchen* and *living room/dining room–kitchen,* where the dining room shuttles between conversation and eating as the landing between stair and floor. In the case of the dining room, this shuttling was the sign of a dual reconciliation, between Denis's and Ingrid's families, and between this reconciled Family Wood and the house in which it dwells. The shuttling of the landing is another sign of reconciliation, but here a simpler one, between no more than the

Family and the house. Nevertheless, in common with all such loci of reconciliation, the landing remains a precious site, not for the sake of the reconciliation (which is radically artificial) but that of the rupture (it is a kind of domestic hierophany). It is in these places alone that the *house* retains the distinctiveness of the social competence that precipitated it, that the uniqueness of *its* voice can be distinguished among the clamor of the family attempting to absorb, to engulf, to *inhale* it: the house *will be heard.*

Though not just in the debate between the rules: it will also be heard through the speakers in Denis's study. The landing is anomalous not only in its shuttling between stair and floor; it is also unique among the room's *extensions* (via the floor into the dining room, through the door and windows onto the porch), in the way the pseudopod of the stair lifts the landing up, out, and *over* another room—as though the poop deck of a ship—*over* Denis's study, over the turntable of the stereo system in fact, so that any kid bounding onto the landing from the second floor is sure to cause the needle to . . . skip maddeningly, to cause Denis to thunder "GOD DAMNIT NO RUNNING ON THE STAIRS," to cause him (finally) to bring the kids into his study while he bounds down the stairs allowing them to hear for the first time the proximal sound of a distant bound—"Oh, we didn't know . . . ." The *t* in the k*t*PRO, consequently, refers not to the landing, but to the record on the turntable in a distant room (it is down the stairs, through the living room, and around a corner of the dining room), even in RULE 66 where the record isn't mentioned, though here it cannot be overlooked that this lacuna gives it the general form of other stair rules *(don't play, don't slide, don't kick, don't run, don't clomp).* By evading this form, Chandler's version is enabled to express the etiology of the rule: this was: that the house *was being heard* . . . where Denis didn't care to hear it.

The Stair

Christmas morning. Quiet. Until *pwop:* one of the kids' feet hits the floor. *Shuffle, shuffle, creak, plap:* the toilet seat hits the tank. The sound of the pee is like a garden hose in a half-filled bucket. *Shuffle, shuffle, creak, flop:* back to bed. Silence. Then . . . *tinkle, tinkle, tinkle:* the bells on a stocking, urgent whispers, another voice now, *pwop, shuffle, shuffle, creak, plap:* the sound of the pee this time is like a fire hydrant opened on the street. *Shuffle, shuffle, creak, flop:* back to *his* bed. His *whisper, jingle, jingle, whisper.* Real words raised, then hushed. Silence. The low slow stridor of a door being *eased* open. The long *squeal* of a tread slowly taking weight. Another. Another. From the landing, high smothered exclamations, then the quick *squeaks* of retreating feet.

The passage of the stair cannot be hazarded now, not alone anyway, not without Denis and Ingrid. Santa's visit has changed the relationship of upstairs to down, transformed that between the top of the stairs and the living room, this now the site of the tree, site of the Visit, site from which stockings were removed to the bedrooms upstairs, Santa's feet on the way up igniting the treads, leaving the stairway a cataract of flame. Too hot for little feet: *quick! Back to bed!*

The flame plunges eleven feet from the nosing of the second floor where this juts out over the nearest tread, eleven feet in a run of sixteen. Unless it is recalled that this includes the landing, it sounds more generous than it is, but in fact the stair is canonical in its proportions. *Using* the stair it's difficult to appreciate it for the simple machine it is, an inclined plane taking the turn into a screw cranked up through the nub of the house, a simple machine of oxymoronic complication, of—even at the level of the rules—no less than five parts and each of these a thing of many others, the seventy-three visible pieces of the newel post, the seventeen risers, the fifteen treads, the banister, the twenty-seven balusters, the nineteen floor boards of the landing, the stringers, the stud partition below the closed string of the balustrade, the moldings tossed everywhere like confetti. Even in its fabrication it is nearly canonical, except that the stringers run up beneath the treads, and the risers vary in height, and they aren't tongue-and-grooved to the treads. But otherwise it's the light white stairway with blond oak treads it's supposed to be, pulled back by the landing to make a space at the foot of the stairs to swing the door and still have space to stand in, decorative yet practical, linking the public and the private worlds, a way to get furniture upstairs, and dirty clothes down.

And to sneak down at Christmastime for an illicit peak at what's beneath the tree.

rules   70. "No playing on the stairs" (Denis, ktcON, kPRO).

71. "No running on the stairs, it makes too much noise" (Randall, kAPP, ktPRO).

72. "No clomping on the stairs" (Denis, kAPP, ktPRO).

73. "Stairs are for going up and down" (Ingrid, ktcON).

The function of the stair is certain (it is for going up and down). But what is its meaning? The glance links it to the door (both are wooden contraptions painted a glossy white) and the floor (with which it shares a blond oak surface finished as glass). But the rules denote a subtle severance. If we take for the door the syntagm *screen door* (1), *door* (2), *doorframe* (3), *window in door* (4), *windows in lights* (5), *bells* (6), and *lock* (7), then the door is the object of forty-one rules. The preponderance of these (though not quite half) are concerned with appearance, the minority (just less than a third) with protection (some two-fifths with control). The floor (8) is more extreme. Nine of its ten rules have an interest in appearance, three in protection (four in control). The syntagm *stair* is different. Here appearance is ignored (the orientation of no more than four of nineteen rules) despite the evident self-identification of the stair with the alliance of the white and the bright, the beveled and the glossy. We have seen moreover that even these few appearance rules are essentially empty *(don't kick, mar, the risers)*, concessions to a taxonomy of *substance* (the risers are wall) over that of *form* (the risers are shattered wall, they belong to Chaos). Since as substance the stair is without independence, to treat it as such is to miss entirely what is stair. From the perspective of substance the stair is only *wall* (risers), *floor* (treads), *woodwork* (balustrade). But what is *stair* is precisely the way this wall *recedes before you* as you climb, the way this floor (something you can stand on) *gives way before you* (is something you can tumble down). What *is stair* is the way this wall is *not* wall, this floor *not* floor, is this twist, this *is* and *is not,* this subversion *in* the room of what would seem to be defining *of* the room, this smashing and scattering of walls and floor. The result of this cataclysm is that the room is opened out (the walls are broken down, the floor—the ceiling— is broken through), is that the room is connected to other rooms, to the world (is permitted to be an organ, a functional structure in the organization of the home). In this work it joins the doors and windows, but where these are framed *absences* (where the frame celebrates but *contains* the emptiness that is their essence), the stair is an unframed *presence,* a monstrous contrivance seeping through balusters and leaping banisters, protruding into the room even as it leaks the room . . . out . . . somewhere else (who knows where?): it is dark at the top of the stairs.

It is always dark at the top of the stair. The keynote of the stair is inevitably: danger. Though many of these dangers cannot be defended against by treading carefully—the way, for example, it subverts the categories of floor and wall, or the treacherousness of the passage from *anonyme* to *intime* (though even these sooner or later converge on *falling,* falling apart, falling in and out of love), the stair nevertheless carves out a place beneath the aegis of the protection code in sharp contrast to door and floor. Nor is it that *these* are free of dangers (recall no more than the hysteria evoked by the memory, or pseudomemory, of a finger being crushed by a door), but that the stair presents itself as a site of many dangers *recurrently encountered,* each modeled on the archetypic fall, origin of the truest vertigo, sign of the ultimate abandon (in falling, one is . . . lost). The stair will be approached with circumspection or like the flouted god it will trip us, maim us, kill us (isn't this the fear?).

Yet the obeisance shown by the rules to safety at the expense of appearance (most *visible* in the toe talk on the risers) does not manage to establish the stair in paradigmatic opposition to the door. This is because whereas circumspection with respect to the stair *is necessary* (it really is dangerous), the stair itself *is not* (it shares with the doorframe a notable superfluity). The dangers it imposes on the family are risks *the house* accepts in order to be two-storied, this in turn a sign whose mobilization is essential in a home for an aspirant to the upper classes (this was especially true when the house was built). Nothing else so effectively marks his difference from those unable to afford two-story housing (determinatively those in the shotgun houses—since bulldozed—of Raleigh's black Southside) as well as his similarity to those up the street in Montfort Hall (the Italianate mansion built by William Boylan when Boylan Heights was still his plantation). Inner sign of outer glory, the stair sharpens *within* the distinction between public and private (domestic social space is stretched), as the two-story house *without* sharpens an analogous distinction between lower and middle classes, between black and white (urban social space is stretched). These distances can be summed: we may add (1), the distance between the Southside shotgun and the Boylan Heights house, to (2), that between the living room and the interior of a bedroom on the second floor. This new distance is immeasurable (it is that between the bed of the slave and that of his mistress). To cross this distance remains in the South a transgression of unspeakable enormity. Is it any wonder that the stair, principal cipher in each addend, protrudes into the room, that its balustrade gleams, that the newel post bristles? That it is not a ladder jammed through a hole in the ceiling of the kitchen?

In the urgent necessity *(I'll be right back down: I have to go upstairs for a second)* the sense of personal responsibility is overlooked, but people choose to live in two-story houses despite their inconveniences, choose to mobilize these signs despite their chilling cost. Do not year after year stairs top the list of causes of accidents within the home? Are not infants kept from the perils of the stair by temporary gates? Do not old people flee them as soon as the cost of the climb exceeds the value of the sign? That it would be impossible to build the floor space of the house on the area it occupies without going up goes without saying (this is the alibi of the necessary), but neither the volume of the house nor the area it occupies is innocent (and why not go down or spread out? the house does not *fill* its lot). Both are no more than *other signs,* kin to those of the *second* floor (the very ordinal notes the excess), the quarrels in the sidelights, the Georgian Revival door, the beveled glass, the glassy floor . . . the very *idea* of the private house.

Within the room, especially one scarred by the violence of a stair, the history of the private house cannot be escaped, neither that of this particular house (built by lower-class blacks to make a home for white aspirants to upward mobility), nor that of the private house in general, a history from its inception of increasing social distance, of ever more virulent privatization (the private house is the tombstone on the grave of the face-to-face community). Although it predates the emergence of a bourgeoisie, the private house will be seized by the bourgeois as the primary affirmation and the physical manifestation of their individualistic and competitive values. In their hands the private house will become what it is today, no more or less than the form in the built environment of: *the nuclear family* (the single family home), *sexual discipline* (the single family home, separate bedrooms), *ambition* (to own your own home), *measurable accomplishment* (owning your own home), *loyalty* (to the family, to the home), *respect* (for our wives and daughters, for our homes, for our fine neighborhoods of homes). To encounter a private house is to experience these values, *no less wrapped up in industrial capitalism that the house itself,* that contrivance of contrivances that in its unceasing division of space (into closets, corridors, rooms, floors; into yards, lots, blocks, subdivisions) makes possible the living of these isolating and emulous values. All this slides into the room down the banister of the stair.

But ours is a room that declares its openness (there are no curtains on the windows, the door is unlocked), that will argue for a kind of community (portraits of anarchists and revolutionaries), that is diffident about the very forms of bourgeois life it develops, expresses, exhibits (this critique lies at the very heart of the modernism the room espouses). Despite this, the rules Denis and Ingrid have emitted for the doorframe and floor acknowledge their bourgeois origins (the room is an expression of values at war). Why shouldn't these that cover the stair?

Because they don't have to. The danger of the stair at once excuses the eruption of rules falling under the protection code *and* preempts the need for rules devoted to appearances. After all, rules emerge only to protect *threatened meanings:* kids cannot threaten the meaning—cannot sully the bright whiteness—of a stair they may not touch (tAPP); nor exhibit the abandoned attitudes of the animal—sensuous sprawlings, sinuous slitherings—on a stair on which they may not stay (kAPP). Appearances here are preserved (and control achieved) by *transcendent prohibitions* (their immediate object is safety). Yet the stair does, in extreme degree (that of potential concussion), emphasize domestic and urban alienations which Denis and Ingrid cannot fully support. As limited as this distancing is, the missing appearance rules *do* signify. Apparently the stair is another of those sites—the dining room, the landing of the stair—where the house will be heard, where the values of the competence that constructed it and those of the family that dwell in it diverge.

A room in which this split is incarnated is self-conscious (it is poetic: it draws attention to its signifiers). It makes self-evident its being as *archeological site* (in its preservation of values of turn-of-the-century Raleigh), as *economy* (the purchase of this house by this family exemplifies the dictum that though we make our own history, we do not make it exactly as we choose), as *performance* (the room variously throws its weight behind the implications of privacy insisted upon by the stair and those of openness in the unlocked door, the uncurtained windows), as *interpretation* (in the reconciliation articulated through the living that *is* the room). It is this precious confusion, this dramatic doubt that the stair rules foment: as encountered in the rules, the stair stands neither with nor against the door, but . . . beside it.

XIV · COLLABORATION

We have seen that the field of rules comprises a spectacle of rationality. The kids will confirm this, they will insist upon it, they will illustrate it. For example, Chandler will justify Denis's *don't jump on the landing* (RULE 66) with *because the record will scratch* (RULE 67). In this way the child participates fully—as did his parents before him, and theirs before them—in the elaboration of a system initially no more than imposed. But such imposition has no more chance of success—of "taking"—than the occupation of a hostile state. Therefore, the child will be a collaborator . . . from the very beginning. That the rules are arbitrary impositions will be repressed, both forgotten and forgotten that it has been forgotten (not only will there be no awareness of arbitrary imposition, but of even this lack of awareness there will be no awareness).

The child a collaborator . . . And why not? The home is alive, it works: *the trains run on time* (this is the spectacle of functionality). The child represses its knowledge of the origin of the rules because it has been induced to believe that the home works because the rules are followed (this may constitute a tautology); that the spectacle of functionality is itself no more than a function of the spectacle of rationality; that this rationality (this liberal reasonableness) is the very frame of the family (it *is* the family, shorn of flesh and history). If, given a rule, a younger child asks *why?*, older children are able not only to respond with conviction to the riposte *why do you think?*, but, lacking an explanation, to invent one.

This explanation is rarely unrelated to other expressions of rationality in the child's world (embodied in television, comics, church, schools): each rule will be accepted as either (1) the conclusion of a syllogism, or (2) the articulation of a previously un-suspected chain of cause and effect, or (3) a datum of existence. A rule will be an instance of logical thinking (it will be Logic), it will be a lesson in Physics, it will be an item of Natural History. These relays will refer the regulation of behavior out of the domain of the family and into that of the law—that of thought, of matter, of living systems.

Now, each of these is capable of taking refuge behind any other in another instance of that circularity which as we have seen seals off inquiry at a certain level: nature is logical (or vice versa) and living systems are natural, so they are logical (it is by studying them that we learn what logic is), et cetera, et cetera. Instead of being something *we* shape, behavior turns out to be something shaped by natural laws and forces (if only we would allow ourselves to be guided by them: astrology, psychology). This has as one effect that of submerging the behavior of the family in that of class (as invisible as water)—that is, in the behavior of similarly behaved others (confirming its naturalism: *you never see the so-and-sos behaving like that!*). This class behavior in turn is variously referred to still larger "classes" (to Raleigh, to the South, to society), and so on, until that class is reached called "humanity" (that is, human nature). But of this no one has any but the most particularistic experience, so immediately it is referred back down to the behavior of individuals, inevitably, that is, to the level of the family. In this way the child is enabled to acquire the confirmation of an anthropology (knowledge of human nature) without having engaged even *examples* of other systems of behavior. The family is thereby certified as the only form of human organization meaningfully operant on its level: the *natural* form of social organization is the family (which naturally lives in a white, two-story, single-family house).

The necessity of this sealing off of the family from the force of examples of alternative social structures requires that the child annihilate the memory of his or her induction into it, an induction effected through the medium of its collaboration in the articulation of the field of rules (the annihilation of memory is how the sealing off is actually achieved). Origination *in time* will be exchanged for a kind of sempiternity: the rules (that are also the family) always were and always will be (it is through such an operation that the nouveau riche transform themselves into "old money"). The repression naturalizes the *origin* of the rules as it naturalizes the rules. The rules appear not of a family or social order, but of the order of ontology, science, logic.

Exactly. It will be a matter of Logic. Each rule will be accepted as nothing other than the conclusion of a syllogism whose premises will be either "self-evident," or other rules (it will be soritical), or suppressed (it will be an enthymeme). For instance, when asked about rules for the door (2), Chandler volunteered *don't open it too wide because it scratches the floor* (RULE 16). Now this statement of the rule constitutes an enthymeme of the following form. PREMISE: *The door will scratch the floor if opened too wide.* SUPPRESSED PREMISE: *We don't scratch the floor* (this also happens also to be RULE 44). CONCLUSION: *Don't open the door too wide.* By exhibiting the major premise in the utterance of the conclusion (as here in RULE 16), Chandler displays the prohibition as logical, hence: necessary. This makes the rule self-justifying. This in turn makes it possible for him the more readily to embrace it. By masking or muting its arbitrariness (by displacing its arbitrariness onto RULE 44 [whence it will be displaced elsewhere]), the syllogistic form helps erode any resistance Chandler might experience because the rule interferes with his pleasure, or because it reduces his autonomy, or because it subjects him arbitrarily—*why me?*—to the rule of another (his father). When such a justification is not in evidence (has not been provided in the initiatory unfolding of the rule), Chandler and Randall are capable of creating one. Thus, where Ingrid says *pick up things you leave on the floor* (RULE 47), Chandler amplifies this into the consequential *don't leave stuff on the floor so people can trip, like marbles* (RULE 48). This substitutes a discursive rationalization for an arbitrary command and so conducts the rule into the *family of reasonableness* constituted of a field of rules *necessarily* exhibiting reasonableness, that is, coherence, consistency, hierarchic integration. It therefore comes as no surprise to find Chandler impatiently commenting of *there's no leaving of shoes around in this house* (RULE 49) that *that's leaving stuff,* where the impatience derives from his awareness that the rule is logically not required since shoes are certainly among the *things* of *pick up things you leave on the floor* (RULE 47) and the *stuff* of *don't leave stuff on the floor so people can trip, like marbles* (RULE 48). Or again, of *don't use the door as a "gate" in a game* (RULE 12), *don't get your fingers caught in the door* (RULE 13), and *don't hang on the door* (RULE 14) that *that's basically playing with the door,* that is, already implied in the more general *don't play with the door* (RULE 11). This is no longer justification: it is formalization. And Chandler is no longer the simple subject of these rules, but at the same time their interpreter. Implicit is an understanding of the metarules that structure the field (rules of consistency, of hierarchic subordination) which, in intimating, he conspires to create. No longer imposed upon, Chandler becomes a font of rules: *he* will teach *us* a lesson in logic.

Or, he will teach us a lesson in physics. Each rule will be articulated as a generalization of observations, as a law of Science. The form will often remain syllogistic (the logic of science), but it will be the relevance of the rule to an effect not sensible in the prohibited cause that will be displayed. Hereby the rule will be rationalized *(you'd naturally follow the rule if you understood the effect of what you were doing),* even as it will slip from the imperative to the indicative: instead of demanding, it will no more than point. For instance, Randall will explain that we *don't slam the door* (RULE 10) because *you'll ruin the woodwork.* Consistent, he will repeat this apropos *when you close the door, turn the handle* (RULE 41): *yes,* he notes, or *you'll ruin the woodwork.* This calm and apparent acquiescence masks an act of appropriation disguised as a substitution: Randall's sagas of cause and effect shift *the slamming of the door* from being an effect of Randall's agency (*Randall slams door,* as implied in RULES 10 and 41), to being the cause of the chipped paint *(slamming chips paint).* What is swept away in Randall's concern for the physical condition of the woodwork is Denis's concern for the social behavior of Randall, *is Randall,* who as subject has vanished from the scene. Where Denis had been speaking *to* Randall, miraculously they now speak *together* of an objectified set of events in which neither is particularly implicated. The rule does not cease vibrating in the air, but it is no longer especially Denis's: *Randall has appropriated it,* and if he refrains from claiming it his own, he admits it adherent to no other will. It has become . . . a law of science. Another example: Denis says *no running on the stairs* (RULE 71). *Yes,* says Randall, *it makes too much noise.* Again: everything is objectified: Randall and Denis—*two observers of the scene.* A final example: *don't stand on the seat* of the Sof-Tech chair (RULE 126). Randall agrees, observing (pedantically) that you can *fall more easily than on a normal chair.* The rule—*hey, presto!*—is turned into a gloss on the chair. No less smoothly the student disappears. Randall returns . . . as a teacher! *He* will instruct *us* about the laws of science.

The kids *will* make the rules theirs, even those whose authoritarian content is least capable of being reduced (to logic, to physics). These rules the kids will take as data of a frankly empirical natural history (it will be that of *Homo sapiens,* late twentieth century, American, professional). Be the rule ever so arbitrary, it will be integrated into their well-ordered universe as are the no less arbitrary facts in the almanac, in Ripley's *Believe It or Not,* on the back page of *National Geographic World* ("A string mop with a mouth? No, this hairy animal is a medium-size Hungarian sheep dog called a puli.") These are those rules the kids can express to their friends only in raised voices ("No, Kelly! No!"), the ones that test their collaborator's mettle, for here no reason is advanced for anything: *it just is.* For instance, *don't play with the bells* (RULE 36) was probably not given to the kids as necessary to maintain the silence which alone allowed the bells to sign the opening of the door *(sound/silence),* but in the rabid irritation—STOP PLAYING WITH THOSE %&!@!!# BELLS—better caught (the rules are like amber) in Randall's *keep your friends from ding-donging them too much* (RULE 38) and Chandler's *don't ding-dong them too much when someone's in the living room* (RULE 37). Though both of these represent attempts at recuperation, they fall shy of those achieved for rules whose authoritarianism lies less close to the surface. Thus Chandler's simple conditional fails to reach the syllogistic verve of *don't leave stuff on it so people can trip, like marbles* (RULE 48), which might here have resulted in something like *don't ding-dong them so people will go crazy.* And while Randall does substitute his friends for himself as the subject of the rule, he does not manage the effect-to-cause shift that turns an imperative into an indicative (which might have produced *ding-donging the bells irritates some people*). Though both boys brandish these pseudo-recuperative forms, the authoritarian substance of these rules remains untouched: in the end they just accept them. *I got in trouble* Randall says of *don't play around the banister* (RULE 61). And after volunteering *don't tilt the shade* (RULE 149), he notes with resignation that *Denis does it a lot.* Canonically, eating with the elbows on the table is proscribed. When Denis (nevertheless) does it, Chandler makes him take them off.

The rules are theirs.

How else could it be? The family, shorn of the anecdotage of flesh, is no more than the rules that bind it, is no more than those through whom the rules pass, from generation to generation, from Erich and Gerda to Ingrid, from Nancy and Jasper to Denis, from Denis and Ingrid to Randall, to Chandler, from them to . . .

Each in turn *makes* the rules (theirs) in the very process of *passing them on,* claims them only to lay them on another, expending in a spasm which exhausts it the authority originally stolen in a kind of ceaseless *you can't fire me, I quit* down through the ages.

At nine and eleven, somewhere between receiving and laying down the law (though no child—however young—is completely free of complicity; no parent—however old—completely lost to innocence), we catch Chandler and Randall . . . *en voyage.*

XV · CULTURE

The rules are theirs.

What does this mean? It means that Randall and Chandler deploy the rules, insist that their friends obey them, hold each other—and Denis and Ingrid—to them. It means that they create rules (one or the other will say, "I think we should have a rule which says . . . "). Does this mean that families are not the factories, slaughterhouses, prisons R. D. Laing has insisted they are? Not at all. It no more than acknowledges that as in all such organizations the victims collaborate in their situation *(the trains run on time).*

There is no question that initially the rules *are* imposed (anthropology guarantees this). Yet there is something perversely reductionistic in characterizing as no more than oppressive, repressive, and so on an organization—like that of the family, like that of school—in which those upon whom an entire system of behaviors has been imposed against great resistance (boredom, dropping out, refusal, truancy, rebellion, whining, withdrawal, hostility), turn around and impose it on others (their children).

In fact we have a name for the whole system: it is culture.

Culture is a machine for teaching rules that have been taught. The machine subsists in the rules (it exists nowhere else). There is a sense, then, in which the rules are self-perpetuating. In this sense they are not the kids' (they no more than pass through the kids [to their kids]). But neither were they Denis and Ingrid's, Nancy and Jasper's, Erich and Gerda's. Though we often say, "We have a rule which . . . " it would be more accurate to say, "A rule has us . . . "

Yet the rule has no existence independent of those who utter it (culture is an epiphenomenon of human being [but what is human being void of culture?]). In this sense the rules *are* ours. How rarely do any come together—in families or other ensembles—with perfectly congruent systems of rules? In these collisions some rules are reinforced, others canceled out, though in the new regulatory environment most no more than acquire new valences. In novel circumstances, whole new rules may be emitted (this must happen frequently in modern societies).

The system of rules that is culture (that is human being) is adaptive; and natural selection operates to sort the rules, prompting our forgetting or emitting, repeating or repressing. Does this mean that the rules are unconscious? Not at all. It no more than acknowledges that consciousness too is cultural.

The Walls

At first you don't see the cracks, but after you've noticed one you see them everywhere, seaming the walls like lightning. Some crackle surficially from floor to ceiling along the edges of a wallpaper obscured by paint; but others like forked lightning slash through the plaster to the lath below, far below, for the lath is like a cartilage encased in skin.

The surface everywhere—except where the cracks expose the bathic substance—is paint, a millimeter-thick film of titanium dioxide and silicate particles in a matrix of vinyl acetate/acrylic resin. When Denis and Ingrid moved into this house in 1976, they rolled two gallons of Glidden's best interior white latex onto the walls of this room, at that time a light green that had been brushed onto a pre-existing wallpaper (except on the wainscoting by the stairs), that had in turn been glued on top of wallpaper that had itself been glued to the finish plaster.

Yet this finish plaster is not itself basal, is in fact no more than another coat, this one troweled over a substrate of two tons of gypsum plaster and plasterer's sand (there are 390 square feet of plastered wall in the room), squished through the gaps in the meshwork of lathing devised to hold the plaster in place. These short thin wooden strips—nearly 1,400 linear feet in four-foot lengths—are nailed to studs in staggered groups of four: the off-placement and short length endow them with a certain (limited) independence. This is supposed to let the frame of the wall move a little without cracking the plaster.

Along the lower half of the wall up the stair (15), behind the radiator at the foot of the stair (9), and on the west wall of the room to the south of the doorframe (3), the plaster has been dispensed with. Here 104 beaded boards covered by a wide cap molding (it is actually window-stool stock) march up the stairs from the doorframe.

These count for another 85 square feet of wall, depending on the amount chalked up to the room, the amount chalked up to the upstairs hall. Plaster and wainscoting together come to a hundredth of an acre. The lot the house sits on is only sixteen times as large.

All of this is fastened to two-by-four studs on sixteen-inch centers, snugly toenailed to the sole and top plates bounding floor (8) and ceiling (69). Braced by horizontal firestops of further two-by-fours, the studs constitute a (cruder) meshwork of their own (within the still cruder meshwork of the rooms). Here, in the air space between the plaster-laden lath of the inside, and the sheathing beneath the clapboard (underneath the aluminum siding) on the outside, is the wiring connecting wall switches (70) to the lights, baseboard plugs to their power source, doorbell push to the electric chime it operates. Gaps between wall and ceiling are hidden by a crown molding (cyma reversa), as those between wall and floor by a basemold (cyma recta surmounting a double torus) over a baseboard. (The shoe molding belongs to the floor [*it* is sanded and varnished].)

In an old house like this, partition walls come to be as structural as those intended to be bearing, everything shifts, the parts have shuddered together over the years, and in even the most carefully laid plaster cracks appear. They are the scars of environmental battles lost by the house, to the ceaseless loading of the wind, to the merciless pounding from the vibrations set in motion by the passing traffic, to the expansion and shrinkage resultant from every alteration of temperature and humidity. The play of cracks is a surficial sign of these infrastructural realities, otherwise obscured by the lath and the plaster, the wallpaper and the paint.

And after a while, you even stop *seeing* the cracks.

rules   74. "Don't put your hands on the walls" (Chandler, tapp).

75. "Don't mark up the walls" (Denis, ktapp).

76. "Be careful of the walls" (Denis, tapp).

77. "Don't kick them" (Chandler, ktapp).

78. "Denis or Ingrid hang things on the walls" (Ingrid, tapp, kcon, tpro).

What are the walls that one should not put one's hands on them?

They are that of the room pierced by the openings, by the arch into the dining room, by the fireplace and its mantle, by the windows with their quarrels, by the doorframe with its muntins. They are the garment through which these are the holes for trunk, arms, neck. If the doorframe is a lace collar, the walls are the sumptuous cloth of the coat.

Only they are not sumptuous. They are plain. They are white. (They are cracked.) Originally they were papered. We do not know what this was like in the living room, but the ceilings throughout the house were dropped between nine and twenty inches at some time in the past, and where you crawl up into the attic you can look across the dropped ceiling to an earlier wall. This is papered in a subtle stripe sprinkled with pale flowers. Two other papers are visible in the "unfinished" bathroom on the second floor (though seven have been pasted on top of one another): a light blue spangled with bluer daisies and white lilies; a pink spotted with roses and vague butterflies. This interior is easy to reconstruct (it was that of Denis's grandparents): windows—treated as decorative objects rather than piercings of the wall—were slathered with blinds, sheers, valences, and drapes in coordinated stripes and florals. On the floor a generous carpet was worked in a floral pattern. The stuffed side chairs, armchairs, and couch were covered in stripes or florals. Between the woodwork stretched a floral paper. Perhaps a pair of sconces or a mirror in a gilded frame hung on a wall, but "things" were not needed to decorate a room conceived of as a decorative object from start to finish (it was, precisely, a bouquet).

The network of oppositions mobilized here is simple: in distinction to the *corrugations* of cave (barbaric, that is, natural) and cabin (poor, crude, that is, barbaric, that is, natural), the walls of the bourgeois house are *smooth* (civilized, possessed of means, that is, of culture). Hence, especially earlier in the century, they are plastered (there are muntins in the sidelights, there is a bristling newel post). From the plaster four routes converge on its papering. By itself plaster retains too evidently the mark of the natural. It is . . . plaster (it is gypsum). Then, inevitably, it . . . cracks. This *undoes* the very work of civilizing that the plastering performed (it marks the erosion of the smooth, the erosion of *civilization*). Therefore either the walls must be replastered, or the plaster must be covered up. Then again, plaster is readily dirtied (returned to nature) but unlike enameled woodwork hard to clean (there is also more of it). Where conspicuous cleanliness may be readily achieved on glass (and its allies: enamel paint, varnish, waxed finishes, crystal, silver), on acres of plaster it demands a labor that is of an *animal order* (it is that of the mule). Hence only among the working classes do we find that *daily* washing of the stoop or even, as Denis and Ingrid observed in a housing project in rural Puerto Rico, *the walls.* Finally, plaster is too plain. Either it bespeaks a sort of Shaker simplicity, a utopianism at odds with the solid comforts of the pragmatic (realistic) bourgeoisie (it bespeaks an elitist *conspicuous austerity*); or it bespeaks a lack of means (it bespeaks poverty). Consequently any movement from plain plaster is invariably a demonstration of middle-class means, a movement toward bourgeois culture. Painting and stenciling exhibit both (at the very least the plaster is decently covered); but the papered room permits the most lavish expression of taste and means (the paper is . . . a tapestry).

The room as decorated object is the bourgeois artifact in its fullest flower. To uproot this flower it is necessary to strip the paper from the walls (or—even greater transgression—paint the paper white: here the tone is frankly assaultive). Each of the homes Denis and Ingrid have lived have had their walls and ceilings painted white, as (in fact) were those in many of the homes lived by their parents. This (surprising) congruence was reached along distinctive routes, that of Ingrid's parents—migrating from Germany during the early years of the Depression—along a sort of neo-Biedermieresque German-American Art Deco; that of Denis's parents—at home here—along a more doctrinaire but no less imported Franco-German Modernism (Le Corbusier, Mies van der Rohe). Floral paper was rejected in either case (if tolerated when it *came with the house:* the home is inevitably the expression of an economy, of a history), though it was the Woods who most rigorously embraced *white walls,* especially given that the paintings they collected (and about which they were deadly serious: Denis's parents met in the art gallery his mother ran with her sister) would, it was fiercely maintained, profit from them (31, 32, 33, 37, 40, 41, 56, 64). The living room for Denis is thus an extension of early memories (the room is a memory) . . . of the pale cream plaster walls of the federal housing project that was his first significant home (dotted with the paintings of Wray Manning, William Sommer, Saul Leiter) . . . of the off-white plaster walls of the two-family house he matured in (covered by the paintings of Fred Mitchell, Roger Hilton, Adja Yunkers). The differences between these walls (canonically Modern) and those of the neighbors (exemplarily *retardataire*) were articulated in Denis's youth by his parents as those between full and empty, between meaning and meaninglessness, between life and death, between—as his father put it in these words written for the gallery from which at the time he was buying heavily—*dizziness* and *decoration:*

I am not a collector of paintings. . . .

When I was a child I whirled about on the grass to make myself dizzy, to break away from the conventions of the world, to bring joy, to become myself for a brief instant. Today I hang paintings on the walls about me to help create that dizziness. . . .

They are all works of art (charts of explorations deep into the unknown, voyages made with full acceptance, and worked on with complete responsibility) and not the work of decorators. Decorators do not provide dizziness. . . .

These paintings are begun by the artist, but never really finished. It is the buyer, the possessor who might possibly finish them. I nourish myself on them in the same way that I devour a beefsteak, or make love. I do daily battle with these paintings. They force you to move the furniture, to hang them on one wall after another, to try to light them. They are a constant annoyance, impossible to live with. They come into the room, intimidate you, ravish you, seduce you. They are alive. This can go on for a few days or many years. But it goes on . . . that is the important thing. They change your life. You become reborn. And finally you finish the job the artist started. You finish the painting . . . and get rid of it. It doesn't, then, really matter if you give it to a museum or throw it in the trash, after all. . . .

In this field energized by *decoration* and *dizziness,* white walls acquire a redoubled force. Not only do they provide a silence in which the dizzying songs may be sung, but they in and of themselves stand in opposition to the decorated. Not art, they nevertheless stand beside art in their repudiation of the empty gesture, their own vacuity paradoxically valorized as full by the empty busyness of the decorated wallpaper to which they stand opposed. Here they come to constitute themselves the leading tone in that Voice of a Dead Modernity so clearly sounded by the room, in the presence of which the whiteness of the newel post and balustrade, the doorframe and door is discovered to be less unambiguous in its significance than might at first have been imagined (the white of an American Colonial or Federalesque Revival turns out to be that of an interior by Mario Botta). What echo is heard in the echopraxia that is the room? That of the Father, certainly . . . but there are so many fathers. It is not just Erich Hansen, it is not just Jasper Wood. It is Loos, it is Rietveld, it is Gropius, it is Ozenfant, it is Le Corbusier ("law of enamel paint of whitewash: suppression of the equivocal"). It is L'Esprit Nouveau with its Lipschitzes, Picassos, and Grises, it is every new spirit, every brushing away of the distracting, the inessential, the bothersome. It is the gesture of *sweeping* the flies from a shoulder or the face, performed with a brush loaded with white. Again and again. The focus of attention, the concentration of intention, are classicizing; the revolution implicit in the sweeping is romantic: the bourgeoisie is zapped from both sides (it will arise again . . . *with white walls*).

Whatever their history (but the room *is* an iconography), these meanings are at the same time no more than *analoga* of what Denis and Ingrid did when they moved into the house. Of what they did even before they moved into the house. *They swept the old house out.* They pulled up the carpet, they ripped up the linoleum, they sanded the floors, they painted the walls white. And then they hung things on them. They infused the house with a new spirit. They made the house . . . new. And though the competence that originally created the house continued to be heard (in the doorframe, in the layout of the rooms, in the landing on the stair), its voice was strangled, muffled, co-opted. The voice that sang was modern: it was . . . theirs.

Whose it will remain. The kids will not sing here (they have their own room) and so *they will keep their hands off.* There will be no sign of dirty fingers, of rubber soles, of pen, pencil, crayon, or markers (the kids will be well behaved [they will not do to the walls what their parents did to them]). As RULE 78 will make explicit, the clean walls will exhibit not only Denis and Ingrid's values (modernist), but their power to exhibit them: *Denis and Ingrid* [are the ones who] *hang things on the walls.* Claiming Le Corbusier's "right to be severe toward those works which serve no purpose, which are superfluous, which are not essential" by banishing the merely decorative (the wallpaper), Denis and Ingrid unavoidably take on the obligation of determining the purposeful, the essential, and the necessary, of figuring out what these things *are* (of figuring out, deciding—*fixing*—what is essential, what has purpose, what *cannot be done without*). It is the conclusions of this (continuing) deliberation that are hung on walls whose whiteness hypervalorizes them, establishes *these* as purposeful (37, 41, 56), as essential (40, 64), as necessary (31, 32, 33). (All that is needed to render decorative the white is to hang on it the superficial, the trivial, the . . . *decorative.* The white becomes . . . *wallpaper.*)

This is the primary teaching (the room *is* a mathesis). The lesson is not, or not primarily, in the collage, in the painting, in the photograph, but in *their being hung on these walls:* it is that, in a life so short and precious, there is no space for the futile, the vapid, the empty.

And hence no hysteria about the cracks seaming the walls like lightning (they attest to the sincerity of Denis and Ingrid's motives [who spend more money on paintings than on getting the walls fixed {the professor is still a student (this is student housing)}]).

No, really!

The Little Window by the Stair

When the kids were young, before they were old enough to see out the front door, they could stand on the radiator by the stair (9) and look out the window. At night it was like looking at a mirror and at Christmas they could see the tree reflected there: if they breathed on the glass they could melt its sharp points of light into fat pools of color, different in every pane.

Although the kids were tall enough to look through only eight of these, there are twenty-six altogether, thirteen in each fixed sash, set as four narrow lights in mitered muntin moldings which dissolve at the top into a field of quarrels. The stool and a deep, molded apron cover all but the bottom half-inch of the lower rail, giving the impression that the lights elaborately latticed above are all but unframed below. A simple molding smooths the transition from the stiles and the upper rail to the mullion, jambs, and header trim. The whole thing is surmounted by a molded cornice

over a high frieze. Less a piercing of the wall than an excuse for pasting up pieces of molding, there are less than six square feet of glazed opening in the seventeen-square-foot extravaganza.

Sooner or later the condensate from the kids' breath evaporates. The sharp lights recrystallize in the night-silvered mirror, and now and then one of them moves across it as the kids watch an airplane fly from sight.

rules   79. "Don't breathe on the windows" (Denis, ktАPP).
80. "Don't wipe off the windows with your hands" (Denis, ktАPP).
81. "Keep your hands/finger off the windows" (Denis, ktАPP).
82. "Don't smudge up the windows" (Denis, ktАPP).
83. "Don't mark up with hands" (Denis, ktАPP).

What are the windows that one should not breathe on them?

As we have seen, they are gems. If the glass in the doorframe is a graduated *rivière,* the little window by the stair is a brooch, it is set with diamonds, it is pinned to a bodice. This is less decoration than diversion. What the gems, what the sumptuous setting (what the eleven square feet of wood molding) hope to obscure is the absence of windows in the wall (16) occasioned by the presence of the stair (15): it hopes to draw attention from the meanness of the house (which has not only no *center-hall stair,* but no *center hall*).

All the windows in the living room (4, 5, 17, 47, 50) bear the mark of the gem. The window in the door (4) displays this as a bevel. The others display it as a field of quarrels . . . but not in equal degree. The difference marks the extent to which they will work (that is, do more than posture). The sidelights (5) (they are not even called windows) will no more than pose (they will literally . . . stand around): these will be the most heavily encrusted with diamonds (they will dissolve into fields of quarrels at top and bottom). The front (47) and side (50) windows will carry quarrels only at the top of their outer sashes. The inner will carry no mark of the gem at all (they will be the ones to do the work [they will let in fresh, keep out cold air]). The little window by the stair will posture (it will not participate in the control functions audible in the Voice of Comfort) and so it will display a field of quarrels; but at the same time it will perform some work (it will admit light through a wall lacking windows) and so it will *not* display quarrels at top and bottom. Thus the little window will be whole according to neither term of the paradigm (it lacks either quarrels at the bottom or an operable sash): this double castration produces a kind of androgyny, the shame of which is actually that of the unpierced wall. First the window accepts this shame as its own. Then it tries to hide it (the molding, the sparkling glass). The feint works only if the gem can capture our attention (can keep us from completing the paradigm in which it is no more than a mediant). At the very least . . . *the glass must gleam.*

Therefore: "Keep your hands/fingers off the window."

The Plant on the Floor in Front of the Easel

Sometimes, just before Ingrid waters it, the peace lily droops and its lower leaves rake their outer faces on the floor like green dust rags. Then, the stems arc back to the pot like basket handles.

The pot is a rolled-rim terracotta vessel six inches across and six inches deep that stands in a shallow terracotta saucer. The saucer sits in a transparent plastic liner intended to protect the floor from errant water, sweated, or flooded by too heavy a hand on the watering can.

After which the turgid stems stand erect, waving their leaves in the air like flags.

rules   84. "Don't hurt the plants" (Denis, ktapp, tpro).

85. "Be careful of the plants" (Ingrid, tpro).

86. "All plants, don't kill or injure—be nice" (Randall, kapp, tpro).

What is it "to hurt something"? It is to wound. It is to inflict pain.

Things that cannot experience pain cannot be hurt. They can only be broken, that is, split into fragments or smashed into smithereens. In this way the form of the rule distinguishes between the pot and the plant, between thing and life.

This extension of the prestige of the living to the plants at once *elevates* the plants from the level of things to that of people (for like a person, *like a kid,* plants can be hurt), and *blurs* the distinction between people and things (for though coded as bodies, plants remain things [they can moved, they can be repotted, they can be . . . thrown out]). While the first effect draws on principles of kinship (plants and people *are* reciprocally metabolic) and so suggests the relevance to the plants of rules articulated to protect kids as well as things—*don't pound on them* (RULE 34), *don't sit on it* (RULE 55), *don't stand on it* (RULE 56), *don't stomp on it* (RULE 67)—the second effect encourages their still wider application through metaphoric amplification (it makes it possible to imagine enjoining the kids from "hurting the painting"). In this way RULES 84, 85, and 86 render the status of every object in the room (at least potentially) ambiguous, for in that they might be capable of sustaining a (metaphoric) wound, all are algesic, that is, alive, and therefore susceptible of being swept up in the rules governing comportment, decorum, and manners (that is, the rules of the appearance code): they engender the possibility of *having to be polite to the table.*

Is this adventitious? Not at all. One of the motivations for bringing plants (of Nature) into the home (of Culture) in the first place must have been to provide objects of nurturance less cathected than other members of the family (than other people), less cathected (and less demanding) even than pets, with respect to which house plants must originally have been no more than novel sedentary forms (and furthermore not implicated in the historic roles of dogs and cats). As objects of nurturance, plants not only absorb energy but reflect its quality: plants thrive, do not do well, or die. Unlike children—who require to be summoned for display (and are liable in any event . . . to disappoint)—the plants are there, reliably radiating identical (if not superior) virtues. Though it may be a question of a green thumb, that is, a touch, a gift, a talent, in either case that the plants survive is inevitably a dual tribute to the residential environment: for green thumb or patient skill, the plants bespeak not only an ability to raise plants, but insofar as the plants survive the kids, an ability to raise kids as well.

For the kids to respect the plants either as living things or as parallel objects of their parents' nurturance (that is, as *vegetal siblings*) is to manifest . . . a concordance of blessings.

The Easel

There is no easel.

This is the first thing to say about it, that the easel has fled—been taken anyhow—to New York.

But when the easel *was* here it stood, a provocation, just inside the door, just across the space through which one would move to the stairs or the couch, or on through to the dining room, in between, that is, the door and the Mexican Ferris wheel. As ordinarily extended it stood seven and a half feet high and took up five square feet of floor, a substantial wooden presence painted green (except on the face of the wings where the box tray had rested when it was painted), awkward on the gleaming floor beside the glossy newel post in its daubs of paint—clotted on the box tray—like a peasant at a salon.

Like a peasant, it worked.

One of the things it did was to support two sheets of Lucite (these made it possible to display a variety of drawings, prints, and photographs). Another was to permit these to be raised or lowered, an action achieved by a hand crank that drove a snail up and down a rack of pins. This rack was secured to short bars, in turn fastened to left and right wings. The entire ensemble—supported by a back stick that permitted

tilting (to prevent glare on the wet oil)—was hinged to a base, to the four stout legs of which castors had been affixed.

It weighed a ton and was almost never moved.

rules

87. "Don't screw around with the easel" (Denis, kAPP, tCON, tPRO).

88. "Don't wind it up and down like we sometimes do" (Chandler, kAPP, tCON, tPRO).

89. "Denis puts the (chooses the) easel items except at Christmas when Ingrid decorates" (Ingrid, ktCON).

The easel is a relic.

Originally it belonged to Denis's mother's father, that is, to Wray Manning. In the 1940s and 1950s when Denis was growing up, his grandfather had been a painter of mark in Cleveland, and in Denis's memories he sits before this easel in his studio— first on the ground floor, then in a very large room in the basement of the big house in Chesterland—painting the still lifes and landscapes of the countryside surrounding Cleveland that won him prize after prize in the annual May Shows of the Cleveland Museum of Art. There is a smell of turpentine and in the background a bookcase crammed with books of art reproductions.

In the early 1960s this grandfather retired to Florida, where in the late 1970s he died. Then Denis's brother Peter, with their aunt Mary and uncle David, "rescued"—from the studio out back of the house still occupied by the latters' stepmother—books and paintings and the easel. Peter, with the van and the commitment to painting (41, 64), assumed the easel, but Mary claimed it, so Peter, agreeing to hang on to it for her took it home with him to New Orleans. Ultimately—penultimately as things turned out— he stored it with Denis and Ingrid, who put it to work beside the stair, between the Mexican Ferris wheel and the door.

From the beginning it was a keepsake, an heirloom (Denis and Ingrid always told people whose easel it had been), a memento, a relict (it was literally a relict of Wray's studio); and it made of the room *a shrine* (it brought the studio into the room), not only—or even especially—to Wray (whose work only occasionally sat on it), but certainly to the values embodied in it that he transmitted through his daughter (who met her husband in the art gallery her father founded [which she ran with her sister Mary]), to Denis; and which Denis, with or without the easel, will transmit . . . to Randall, to Chandler.

What values are these? They are those we have previously encountered in that isolable aspect of *Denis's* family life synopsized as the practice of the arts of contemplation (of books, music, paintings) and which we have seen dominates the room, setting it thereby in opposition to the kitchen, dominated by that aspect of *Ingrid's* family life synopsized as the celebration of the rituals of eating. That this distinction is an operational one is abundantly acknowledged in Ingrid's ascription to Denis (in RULE 89) of full authority over the choice of easel items . . . except at Christmas, a particularization that mobilizes the paradigm *dizziness/decoration* as well. Yet here, in the presence of an ancestral object embodying the (Manning-Wood) tradition (there is more than a little idolatry here), the paradigm is opened to question: the easel speaks in an off-tone, a peculiar mixture of the assaultive and evasive. This is because, in the opposition between *a site of contemplation* (the living room: the whole history of this room assures us of this) and *a site of production* (the kitchen) (the paradigm here is *head/hand*), the easel is on the side of production (originally it was in Wray's studio, not his living room): the opposition between its paint-smeared crudity and the gloss of the gleaming floor and newel post is dramatically . . . assaultive. At the same time, the easel no less thoroughly evades the paradigm by saturating its terms *within the room* (there are paintings, but there is also an easel on which they could have been produced). Yet these tones in turn are muffled: the assaultive is finally a modernist cliché (it is a vocalic in the Voice of a Dead Modernity heard in the room: *it is almost normal*); while the evasive is muted: in the opposition between *production for contemplation* and *production for consumption* (the paradigm here is *head/stomach*) the easel comes down (uneasily perhaps) on the side of contemplation, no matter how much "bread" Wray may have realized from the work at his easel.

It is this contemplative chord that must be heard in the rules. It is not that *don't screw around with the easel* is not concerned to protect from damage anything *on the easel* (tPRO) or to maintain the position carefully achieved by Denis in placing an object on it (tCON), but that it is *more* concerned to inculcate the appropriate *contemplative posture* (kАPP). Thus the rule not only says *don't touch,* but implies . . . *look.* At the same time it mitigates against the animal abandon that is ever the subject of Denis's attention—especially when he is on his hands and knees attempting to recuperate the position lost because of the kids' inability to keep their hands off the crank so invitingly located . . . out front (where Wray could get at it). Then, ignited by the intransigence of the failing machinery, he had been heard to scream *WHAT THE #&%% WERE YOU PLAYING WITH THE EASEL FOR ANYHOW? IT'S TO LOOK AT %&## NOT PLAY WITH!!! WHAT IF YOU'D KNOCKED THE ERNST OFF? THEN THERE'D BE *##%& TO PAY!!* Needless to say, what the kids heard was *don't wind it up and down* ("like we sometimes do").

But in refraining, miraculously they constrain themselves to . . . *no more than look.*

XVI · TO LIVE THE ROOM

We live our life.

This phrase, exemplary in its banality, appears startling when modified to read "we live the room." Yet the difference does no more than acknowledge that the living of a life is not an independence, not a free improvisation on a stage set fortuitously for the performance, but a limited liberty of expression guaranteed by a willingness not only to follow the script but to participate in the ongoing construction of the theater. The life we live is improvisation–acting–play writing–set construction–theater operation indissolvably whole with the environment that simultaneously sustains and is sustained by it.

It is not, then, that our life is lived *and* the room is lived, but that the room *is lived in the living of our life:* the room is an expression of our living, not a part of it, as our body is an expression of our living, and not a part of it. In this way the room is *lifelike,* as, until we die, our bodies are lifelike. This is to say, as is true for our bodies, that the room is an open system, constantly incorporating into itself new material and new energy from . . . outside, from . . . somewhere else, and therefore constantly changing, constantly developing. This change—this development—is, as we have seen, the resultant of mutually interacting *internal* and *external* factors (it is the resultant, that is, of a living), exactly as *our life* is at all times under the influence of mutually interacting environment and genes.

It is important to be clear here: we are not constructing a metaphor. It is not that the living room is *like* a living system—though it is, and though we shall exploit this simile—but that it is an expression of a living system: *we* live the room: it does not live itself.

This does not mean that the room does not get brighter or darker or warmer or colder without our active participation. This happens because even the "room" is an open system caught up in the flux of energies characteristic of its environment. This flux is at the heart of our living the room. Denis and the kids will be reading in the living room as night falls: one of them will reach up . . . and turn on the light (34, 42). Or, during the fall, a night will come when before going to bed Denis will . . . close the windows (47, 50). Later, a day will come when Denis or Ingrid will go down to the basement . . . and turn on the furnace (9, 45, 46). These actions are not the same as, but are similar to, putting on a sweater, or shivering. The differences stem from the reality that the room is neither clothing nor flesh (each has an individual structure); but the similarities arise from the fact that all three—the room, clothing, and flesh—

are identically agents in the ongoing maintenance of organismic integrity. Furthermore, they demonstrate an important truth about the unending change—about the ceaseless development—they illustrate: at each instant this change is *always* a dual function of the present state of the room (it is a reading room) and the room's environment (it is getting dark outside). Subject to this codetermination . . . the lights come on within.

This dynamism by means of which the room conserves its identity (for example, photic or thermal) in an environment undergoing continuous transformation is experienced at many scales. Thus the room undergoes a cycle of heating and cooling on a daily basis, but this daily cycle is itself the subject of an annual cycle. The range, for instance, between high and low is least in August—when it fluctuates between the low-80°s and the mid-90°s—and during the winter—when the heat is thermostatically maintained between 58° and 62°; and the range is greatest in the spring and fall when it can vary as much as thirty or forty degrees a day. This annual cycle is in turn the subject of a history. Not only is the weather caught up in a climate (also the subject of a history), but the means by which the room adapts itself to this history is also the subject of a history, one recorded moreover in the room's "anatomy" and "physiology," in, that is, the "room" (in its degree of insulation, for instance) and its appurtenances (for instance, radiators, double-hung windows, storm doors with screens). So taken for granted is the contrivance of bourgeois comfort that this history is easy to overlook, but central heat has been common for no more than a hundred years, electrical appliances for even less (Edison and Swan perfect the light bulb in 1879, Tesla the fan in 1891, the vacuum cleaner is patented in 1901), while domestic air conditioning—without which life in the South is now scarcely imaginable—is very much a creature of the past thirty or forty years. Epicyclically geared, this history is entrained in the motion of the whirling years, themselves entrained in that of the daily round.

We blur our view of history when we zoom in on the minute, and we diminish the experiential and phenomenal when we snap in a wide-angle lens, but to take any picture—as we did when we inventoried the room—is nonetheless to suspend ourselves simultaneously at every temporal scale. It is unavoidably a time of day in a season of a year in an age of the world. In this way as well, our room is not *the* room, which is a ceaseless relinquishing, a ceaseless replenishment, but a moment in a season of a stage in the room: Bob was there, it was daylight (the lights were not on), it was summer (the screen was in the storm door, the potted orange tree was outside, the rugs were up, and the fan was down from the closet in the kids' room) but it was early in the season (the door was still open in the daytime). It was when Denis had the baseball game (68) on the coffee table (65) and that piece of wire (63) on the mantle (62) and, oh yeah, the easel was still there, by the stair, between the Ferris wheel and the door . . .

Here is what *The Oxford Companion to Art* has to say about that easel: "The studio easel, a nineteenth-century invention, is a heavy piece of furniture which runs on castors or wheels, and serves to impress the clients of portrait painters." *Suddenly it is a wider history.* It ceases being that just of Wray (and his history) or Peter (and his history) or Mary (and her history, who ten years after she claimed it, asked for—and got—the easel) and enters that of The Painter as bourgeois institution; Denis's memories stand revealed for the relays in the reproduction of a social system that they are; and the room is once again plunged into that sociologics—now from another side of the family—from which it can never escape. Because we are open systems (because we observe our grandfathers at work, because we store things for our brothers, because we return things to our aunts), every personal history is inevitably a social history; and because *we* live the room, *every room is a social construction.* The "we" of "we live the room" can be no less than *the entire social structure* which alone is capable of underwriting the system of values which gives meaning to the rules the kids are expected to obey: *the kids do not touch the easel . . .* and the entire history of how to view Art is established in their behavior.

That the reverential attitude before a work of art is an artifice of parental instruction is, as we know, a reality that Culture labors to deny (no one *teaches* people to stand that way [though . . . why not?]). The appropriate posture, the length of gaze, the tone of remark, all this is passed off as Natural, as no more than innately good behavior (only *breeding* shows). The room is a conspirator in this deception. It too tries to pass for Natural: this is the way rooms *are* (heirlooms, paintings, beveled glass). That it is an artificial construction of a history is simply denied ("oh, we've *always* had that"). The arbitrary and deliberate choices that instantiated it . . . *vanish.* With these out of sight it becomes the more possible to insist that the appropriate behaviors consequent upon the room's presence—which educes the rules establishing the behaviors—is therefore also without cause: *it just is* (the shape the world is in is not our fault: this is just . . . *the way things* [Naturally] *are*).

It is the missing object that returns us to history (to the arbitrary distinctions and choices of culture). The absence of the easel constitutes a discrepancy between one inventory of the room (taken on that day in early summer) and another (this one, the one we have taken now); and this discrepancy foregrounds the room as history, as *open to loss* (Mary asks for her easel) *as to gain* (Peter asks to store the easel in the house), that is, as an open system whose identity is maintained in an environment of continual change (Peter offers, Mary requires, the easel) *artificially, on purpose, willfully,* by those who live the room.

It is not the way things *are:* it is the way we have *made* them. This is what it means . . . *to live the room.*

The Max Ernst Lithograph

When going upstairs with the light in the west, it was often hard to see what was on the easel, the reflection of the doorway was that bright on the sheets of plastic.

But at other times, or from other places, it was easy to see what was there, though even then many couldn't see the point, especially the kids' friends. Even with the Muybridge, they wanted to know who the kid *was* and why there were so many copies of him; and if the Yunkers was evidently a poster, still, *what was it supposed to be?* Neither poster nor photo, the Ernst was even less straightforward. It wasn't a painting, but it wasn't a cartoon. *What was it?*

What it was was page 37 of a *livre de peintre* described in these terms by the dealer from whom Denis bought it: "PRÉVERT, JACQUES. (MAX ERNST). *Les Chiens ont soif.* Au Pont des Arts. Paris, 1964. 61 pp. loose, with 2 color etchings & 25 color lithographs by Ernst. 12-⅛ x 17. Illustrated paper covers with color lithograph by Ernst; gray cloth-covered case. (Number 88 of 300 copies. Another chapter in Prévert's canine obsession, which Ernst depicts with carefree felt pen designs, making occasional use of frottage as a corrective. One of the finest late Ernst illustrated books.)"

Two sheets of Lucite, an acrylic thermoplastic produced from petroleum feedstocks in a process rendering the production of glass and steel "natural" in comparison—16 inches by 30—sandwich the lithograph, itself a paper product *(vélin d'Arches pur chiffon)* of extensive manipulation. Run through a letterpress to pick up the text (set in Garamont and Gill by L'Imprimerie Union), it was also run through a lithographic press where it was *pressed* against the stones to which Ernst's drawings had been transferred by Mourlot Frères. The sandwich of paper and acrylic is secured top and bottom by steel "binder clips."

The level of culture here is unbelievable, certified not by the hand of Ernst, but by the presence of technology pioneered by Gutenberg (the letterpress), Garamont (type founding), Senefelder (lithography), Redtenbacher (acrylic acid), Rohm (acrylate synthesis), and others. Dazzled by the lucidity of Ernst's drawing—or perhaps the reflection of the door—it is easy to overlook, or to *look through,* the culture that brings it to the room.

rules   90. "Same as all pictures—don't put your hands on it (same as windows)" (Chandler, ktAPP, tPRO).

91. "Don't knock it off—never been told, but I assume that" (Randall, kAPP, tPRO).

Chandler makes it clear: Ernst = window (= gem).

The Ernst is a gem, and the easel is its setting (it is another brooch), or it is a gem, and the easel is the velvet on which the salesman displays it (the room is a showroom, it is a gallery, it is a museum).

In this way the Ernst is seen to participate in the social and moral economies of the entrance (before which, in its crystal frame, it exhibits itself); and so to align itself with the bevel of the glass (4), the luster of the floor (8), the enamel of the newel post (10). Because fine art prints = Culture, the Ernst fulfills the very promise of the doorframe (= lace = Culture [= wealth and power]). It is true that the easel complicates this reading (it is not a display stand with a brass picture lamp [it really was the site of a production]), but as we have seen this complication is itself complicated and the muffled and muted tones of assault and evasion introduce at most a slight *hesitation* to the participation of the easel-Ernst ensemble in the very thematics of the Culture required of the relay between *lace* and *wealth and power*.

Against this is the resistance of the Ernst itself: it is not a Renoir.

That is, it is not a French Impressionist painting so completely recuperated by bourgeois culture as to be worth millions and millions of dollars to its elite. This is not to say that the work of the surrealist masters is not being recuperated: it is. Surrealist-period Miros, to pick the most egregious example, have become middle-management office *decor* (that is, they no longer provoke *dizziness,* in Denis's father's terms; in the terms of Le Corbusier, they are no longer *essential*). This is inevitable. But against it surrealism raises this defense: it is not an art movement, it is a way of life. Every surrealist has always insisted upon this. "Surrealism," Franklin Rosemont reminds us, "is above all a method of knowledge and a way of life; it is *lived* far more than it is written, or written about, or drawn." Reminding us that for Breton, surrealists were "specialists in revolt," Rosemont defines surrealism as "an unrelenting revolt against a civilization that reduces all human aspirations to market values, religious impostures, universal boredom and misery." And it is an important memory for Denis to recall when he was sixteen sitting in a hotel room in Villahermosa listening to his mother attempting to translate into Spanish for a young man they had fallen in with his father's impassioned definition: *"el surréalismo es el anarchismo del amor verdad."*

What any of this means in practice—the family were tourists in Mexico on income Denis's father had made selling advertising art—has always been deeply problematical for Denis, steeped as he is at the same time in the radical empiricism of the founders of the great tradition of Western science (nor is he even entirely unsympathetic to the positivist program [while sympathetic to Einstein's rejection of it]). Certainly he has never understood how to *live* a revolution. His acknowledgment of the necessity of the proletarian revolution only raises in him guilt for his middle-class habits and tastes (as for the *livres de peintes* of the surrealist masters), while his repugnance for the sanguinary aspects sure to be entailed forces him to reexamine again and again the bankrupt doctrines of capitalist liberalism for any way out. Accepting as applicable to himself Rosemont's characterization ("immobilized beneath a seemingly inflexible net of counterfeit hopes and fears—hopeless and fearless at the same time before a destiny that could hardly be more ruinous to the free development of the human personality—men and women go on fabricating illusory foresights and pitiful after-thoughts as if nothing more important were at stake then the price of cigarettes"), Denis admits that for him, mostly it is . . . *unclear* (the room and its contradictions testify to this).

Be this as it may, Denis's discovery among his father's books of Max Ernst's *Beyond Painting* changed his life (there is something of the relic in the lithograph too). If he could not understand "the anarchism of true love," he could make every kind of sense out of Ernst to whom he still (hyperbolistically) attributes any ability he might have to see and think. The discovery that followed of Ernst's collage novels unleashed in Denis an eruption of collages (first shown in a gallery his father ran), the making of which continues as an important ongoing project in Denis's life (32, 33). The room, thus, even without the Ernst, vibrates to a surrealist string (the Ernst made it more audible). This vibration is out of phase with that set in motion by the circuit of meanings through bevel, floor, and newel post that turn the Ernst into a gem. Against, for example, the competitive and individualistic values we saw manifested in the stair, surrealism opposes its collective and internationalist adventure; where the stair be-speaks the values of the nuclear family and sexual discipline, surrealism, again in Rosemont's words, "aims at nothing less than complete human emancipation, the reconstruction of society governed by the watchword *to each according to his desire*"; where the stair rises for ambition, measurable accomplishment, loyalty, and respect, surrealism invites us to explore the dialectics of eroticism, chance, and the game.

In the room this vibration is often fugitive, but it has its allies. Ernst, the collages (32, 33), and the Ron Rozzelle (37) are linked through a thematics of the dream; Ernst, Peter's box (31), the painting of Malatesta (41), and the painting of Genovevo de la O (64) through a thematics of proletarian revolution; the Ernst, the Lacandon drum (39), and the Oaxaca king (55) through a thematics of "the savage heart"; the Ernst—via the collages—the collages (32, 33), and the wire on the mantle (63) through a thematics of the *trouvaille* (furthermore these are linked with the wicker rocker [35] and many of the records [27] in the thematics of "the found" and "the used"). More generally it is allied to the iconoclasm announced by the bells on the door (6).

These networks introduce into any reading of the Ernst a *frisson*. It *is* a gem (it is literally an *objet de luxe*). But it is also a call to revolution *(les chiens ont soif)*. It is a summons to life (it is a portrait of the Bird Superior, Loplop, Max Ernst's personal phantom). But it is also . . . *hands off!* (it cost more than all the furniture in the room put together).

And the whole room is like this.

XVII · FROM SOMEWHERE ELSE

The networks of meaning do not originate in the room: they participate in those of the world, of the universe. This is because the room is an open system.

We have seen that to inventory the room is to suspend it simultaneously at every temporal scale. This is no less true for space. "Here"—in the case of the room—cannot mean "no more than within the room," but within that nesting of spheres each of which is required for the living of the room. For example, we know that the room is essentially established through its participation in the metabolics and dialectics of house and home (for example, *front/back, mouth/anus*). This, of course, does not mean that the room is the center of the home (the home seems actually to have no center), but it does mean that *to be in the room is to be in the home*. This is not only, or even especially, because the room is *inside* the home, but because everything the room is (and is not) takes the home as its donnée. Identically, because the *home* takes the neighborhood as its donnée, *to be in the room is to be in the neighborhood*. This is uniquely evident when the risk the house accepts in order to be two-storied (and so participate in the neighborhood dialectics of *top of hill/bottom of hill, rich/poor, white/black*) shoves a staircase up the room. Similarly, because the neighborhood takes the city for granted (and so participates in the city dialectics of *inner city/suburb, orthogonal grid/curved streets*), to be in the room is to be in the city, to participate in the life of the city, *to live the city*. Here it is the doorframe with its Georgian revivalist touches, its quarrels in the sidelights, that most completely inhabits these paradigms, but the whole room is unavoidably implicated . . .

. . . in everything. Nothing except *the living* of the room is original. The easel ties it to Cleveland and Florida, to New Orleans and New York. The Ernst links it to France, the Ferris wheel to Mexico. The oil that heats the room propels it into the politics of the Middle East. It is not that the room observes some internationalist program (on

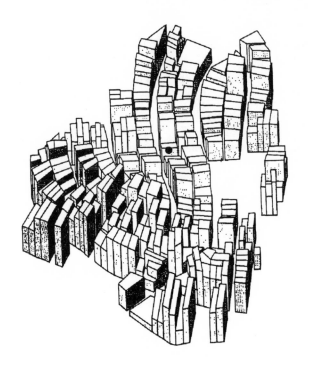

Boylan Heights *(top)*: each oblong indicates, in its plan, the shape of the lot; in its height, the conformance of the house on it to the Boylan Heights norm—a single-story dwelling caught between a shotgun and a bungalow. Tall, thin oblongs indicate the presence of this normative house: small, long lot occupied by a one-story bungalow or shotgun house. The larger the lot, the more distinctive the house, the lower the conformance. The bottom of the hill the neighborhood covers is at the bottom of the illustration. The old plantation house, Montford Hall, is center top. The Woods' house is indicated by the dot.

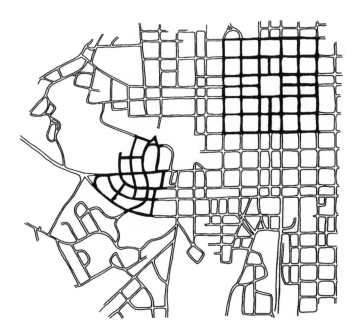

The orthogonal core of old Raleigh *(bottom)*—highlighted upper right—laid out in the late 18th century; the curved streets of an early suburb, Boylan Heights—highlighted left and below—laid out in the early 20th century. In the late 20th century, both are inner city, but the history of their original distinction remains in the street plan.

the other hand, the room does not "Buy American" either), but that the world is tied together with complicated implications of mutual reciprocity at every scale. When at night moonlight bounces off the coffee table (65), it is not the Bauhaus alone that enters, but the moon; and when the sun shines in, it couples the room directly to an energy source 93,000,000 miles away. With starlight, the room links arms with the cosmos.

Because the universe is whole, *here* and *now* (still here, still now) are nevertheless embedded in *everywhere* and *always*. The room and its circuitry of meanings are not exempt: to learn to live the room, therefore, is to learn to live.

The Speaker in the Sewer Pipe

When the easel was in the room, the speaker in the sewer pipe was easy to overlook.

Not that it was notably retiring . . .

A shallow black cone eight inches across, suspended in a cast aluminum frame, the speaker itself wasn't much to see (even the shiny dome at its center didn't draw attention to itself). But the sewer pipe was a sixty-pound cylinder of vitrified clay flared into a ribbed bell at the bottom (where, in its ordinary line of work, another section of pipe just like it would lay its mouth). It sat up from the floor on three evenly spaced alphabet blocks and came via Worcester, Massachusetts, from the AVP Company in Somerville, New Jersey. The speaker was made in Los Angeles by the James B. Lansing Sound Corporation and came to Raleigh via Worcester and Cleveland. It could be silenced by a button-operated switch that protruded from the space between the frame of the speaker and the lip of the pipe.

Twelve inches along the floor and twenty-eight inches high, the sewer pipe with the sun on it was something to see even *behind* the Ernst or the Yunkers or the Manning.

But after the easel left, it rose from the floor like a smokestack from a field of wheat.

rules 92. "Don't touch it" (Denis, tPRO).

93. "Don't play with it" (Randall, kAPP, tPRO).

94. "Don't press the button when something is playing" (Chandler, kAPP, tPRO).

95. "Don't drop things on it from the stairs" (Chandler, kAPP, ktPRO).

96. "Don't play under it—like Transformers" (Chandler, kAPP, tPRO).

97. "Don't put stuff on top of the speaker" (Chandler, kAPP, tPRO).

98. "Cover speaker if you're doing work around it that might allow stuff to fall on it" (Ingrid, tPRO).

From the perspective of the rules, the speaker is a sort of *anti-stair*.

This is because, though both appear under the aegis of the protection code, they do so from opposite sides. That is, the rules that cling to the stair (15) protect the *kids* from it, while those that adhere to the speaker protect *it* from the kids. This difference is rooted in material differences between the stair and the speaker. The former is a crude and inert contrivance of wood and nails. It can hurt and even kill. The latter is an active and sophisticated apparatus of specialized metals and electric circuits. If few people are maimed by their loudspeakers, loudspeakers, especially those invitingly placed face up in sewer pipes twenty-eight inches off the floor, are readily maimed by people, especially kids (the dome, for example, was dented so often that Denis became expert at popping it back out with a piece of tape). Much of this is no more than irritating (Lego blocks bouncing around on the cone), but holes cannot readily be repaired, and reconing is required of one separated from its frame. Because all of this is (at least) annoying, the kids are barred (in any way) from touching the speakers (38, 54), a covering rule in complete contrast to that for the stairs (which was, *please* touch: don't fall). As prophylaxis against annoyance, most of the rules are coded KAPP (because *good kids are not annoying*). But because the speaker (or the amplifier, which is the object protected by RULE 94) really can be seriously damaged, *all* the rules are coded tPRO. This difference in coding distinguishes Denis's and Ingrid's rules (RULE 92, RULE 98) from those of the kids. The parents' rules . . . *protect the speaker*. The kids' rules also protect the speaker, but *by way of* interdicting proscribed behaviors. Thus Randall enjoins *playing with* the speaker (RULE 93), as Chandler enjoins *playing under it*. That the kids' rules address their own behavior (and not the speaker) is aptly illustrated by Chandler's *don't drop things on it from the stairs* (RULE 95), which unavoidably draws attention to the implicit violations of the injunctions *don't play on the stairs* (RULE 63) and *no playing on the stairs* (RULE 70), as well as Ingrid's admonition that *stairs are for going up and down* (RULE 73). At stake in the kids' rules are precisely those tokens of animal abandon (WHAT WERE YOU PLAYING UNDER THE SPEAKER FOR ANYWAY???) ever addressed by rules of the appearance code.

The rules create a further distinction, between the speaker (object of every rule but one) and the pipe (object of none). Impossible to hurt the pipe (even closer to Nature than the stair [it is literally the cooked earth the floor pretends to be]), it is hard to be hurt by it. Nor does it require to be controlled. From the perspective of the rules . . . it doesn't exist. Nor do speaker and pipe form a necessary union. In fact, today the speaker is mounted in an enclosure in Denis's study (it is one of those through which the landing [14] is heard); while the pipe is outside supporting a terracotta saucer trying to pass itself off as a birdbath. Yet they *were* inseparable in the room: pointless apart, together they made beautiful music.

This implies a history (invariably [it will prove to be that of the room]).

Denis was given his first record player when he was in the sixth grade. This was an ancient portable that had been replaced only a couple of years previously by one of his father's Christmas extravaganzas, a Voice of Music console of a vaguely French Provincial character. But much had happened—Richard had become a friend and had built a bespoke system for the family—and so Denis ended up with the suddenly "old" Voice of Music. A couple of years later still, Richard replaced the JBLs originally used in his system with speakers of novel construction (he had glued eggshells to the center of the cones), and Jasper gave Denis the "old" speakers (thus "pushing" the console on down to Denis's brother Christopher). How was Denis to use the speakers? Richard suggested sealing them in stacked sewer pipes (fifty inches high, these were notable indeed). With Dynaco amplifiers and the later addition of a center-channel woofer (in a piece of sewer pipe fifteen inches across), Denis had a better sounding system than any of his friends (and was beginning to acquire the records [23, 26] to go with it). When Denis and Ingrid merged their possessions, this system—initially with sewer pipe newly bought in Worcester—supplanted the old portable Ingrid had had from Dottie Monroe (when she moved to Japan). When Denis's parents "retired" to Mexico, Denis ended up with the now quite old—if often upgraded—system Richard had built, which Denis now merged with the sewer pipe system (augmented by the old Cleveland pipe). This is the phonograph Denis and Ingrid installed in the living room (though a pair of the pipes was really around the corner in the dining room). To forestall the spaghetti of interconnect wires it had been her bane to dust, Ingrid insisted they go through the crawl space beneath the floor. The five holes Denis drilled through the floor in the vicinity of the sewer pipe attest to his novitiate as home owner no less than his incompetence.

Pipes came unstacked. The easel came, and went. Speakers (38, 54) succeeded speakers. While the "fill" provided by the speakers in the pipes had seemed originally to enrich the sound from the other speakers, increasingly it seemed to muddy it. Loath to part with what had been so long with him, Denis first installed a switch to permit him to cut the speakers out (RULE 94 was emitted to protect the amplifier from its unintended use), but finally the excess was admitted, the pipe in the dining room—moved to a window—was filled with a flower pot, and that from the living room—moved outside—was topped with a saucer (though not before we inventoried the room). The trajectory is classical: commencing in want (the pipes were a cheap way to enclose cast-off speakers), they immediately became the site of a precious distinction (they were novel, they were witty, they sounded terrific [they were ruggedly utilitarian {they were *modern*}]), which, after Denis began to work summers on an ore dock, only accumulated density (the sewer pipes were industrial [in the living room they were transgressive]). To the college professor Denis turned into, these meanings seemed less and less certainly embedded in the pipes (especially as other better speakers made them sound less terrific). Finally, as speaker enclosures, they amounted to no more than mnemonic presences (they were never used), evocations of a past (they were precisely . . . mementos) heavy with the nostalgia of regrets for possibilities unrealized. When the grip of these claims weakened, the pipes slipped . . . out of the room.

As the evident consequence of resource constraints, the pipes initially insisted upon the room as economy. As marks of an eccentricity, they energized the room as culture. Toward the end, it was the room as memory they embodied. Now gone, there is nothing but the room which they left (the pipes were *ancestral* to the room that endures), its history (again the absence of the pipes compels us to attend to it), and the five holes Denis drilled to accommodate them.

The Plant on the Record Cabinets Just inside the Door

The deeply incised leaves of the monstera are like large green hands with too many fingers. On their long spathed stems they give the plant an air of a glad-handing scarecrow.

Which is about how much room it takes up. Fifty inches in its longest dimension, it is thirty-four inches deep and twenty-four tall. It doesn't fill it, but this twenty-three cubic foot "waving" volume is about that occupied by, say, Denis and Ingrid with their feet together and their hands by their sides. The plant is rooted in potting soil in a six-inch plastic container tucked away inside a cachepot of white ceramic decorated in low relief with plants and animals of the jungle. This was a Christmas gift from Ingrid's brother, John, and his wife, Bootsie. The monstera came from Christine, a neighbor up the street.

A fleshy yellow green, the spathe sheaths its spadix as a prepuce its glans. When they dry up and fall off, the spathes look like long yellow beards, or circumcised foreskins.

They could scare anyone.

rule   99. "Just like all the other plants—don't abuse it" (Randall, kAPP tPRO).

We have already seen that houseplants are doubly amphibolous: coded as bodies, they remain things; objects of a Cultural production, they are nonetheless of Nature. To these we add a third: evidently clean, they are rooted in dirt, which thus gains sanctioned entry to the home (no muddy feet, but vermiculite in pots). This drift toward Nature is accelerated when the plant in question is Tropical, marked by every variety of overheated excess: the incisions in its leaves are deep (this is its voluptuous-

ness), its habit is expansive (this is its luxury), its temper is unrestrained (this is its passion). These transgressions of the monstera (etymology: evil omen, monster, monstrosity) are aggravated by that of the cachepot whose surface erupts . . . with palms, rampaging elephants, other signs of a Nature *out of control.*

These marks of the amok remain shallow eruptions, they inscribe the surface of a container which holds a container in which the roots of the monstera are unavoidably pot-bound. It is all a dumb show, everything is firmly under control, it is ironic. The violation of Culture by Nature is . . . *mimed.* Can this not be regarded as an instance of the evasive, each term in the paradigm *culture/nature* being here instantiated in the room? Only were the monstera free to break its pot, root in the floor, and climb the walls to the ceiling.

The Small Record Cabinets

Against the white of the walls, the cabinets disappear. There are only the parti-colored records, floating in space.

There are 688 of these, more or less evenly divided among the eight boxes that comprise "the cabinets," each box a seven-faced cube, 13½ inches on a side. This is the modulus of Palaset, an eleven-unit system of boxes, bases, and drawers manufactured from polystyrene in Turku, Finland, by Teston Oy. On the outside of each face a shallow hole has been drilled in the center for the plastic dowel intended to hold the units together. This seems to be working for the six boxes nearest the door (2), but a ⅝-inch gap separates the top of these from the top of the pair nearest the window (47). Once this has been observed—though most people seem not to notice—it is hard not to see the ⅝-inch gap between the floor and the base below this window pair. Because the floor slopes so precipitously here, the individual Palaset bases were dispensed with, and the cubes mounted on a ¾-inch white-painted plywood plank. This lies on the floor beneath the boxes toward the door, but juts out into space beneath the pair by the window, propped up by an "invisible"—unless you're on your hands and knees—block of ¾-inch plywood. Originally this sufficed to close the gap at the top of the cubes, but the seven years they've been in place have seen the front of the house jacked up, footings poured, new foundations constructed and a new sill emplaced to forestall the collapse of the house brutalized by the incessant jackhammering of the traffic.

Polystyrene is prepared from ethylene and benzene, the former derived from natural gas or petroleum, the latter from coal as a by-product from coke ovens, in the presence of a suitable catalyst (like aluminum chloride). Along with, among others,

Switzerland, Cyprus, and the Bahamas, Finland is a country without oil. It lacks coal too. Where does Finland get its petrochemical feedstocks?

In just such a way does the room establish contact with the Urals.

rules 100. "Don't abuse them—stereo used to be there" (Randall, tPRO).

101. "Don't leave anything on the cabinets" (Denis, ktAPP).

102. "Yes, that's a general rule, but the mail is placed there if you can't put it out" (Randall, kAPP, tCON).

Randall is right, of course: the stereo did use to be there (the one Richard built for Nancy and Jasper thirty years ago).

For the kids as well as their parents, then, the room is a memory (this is the only living room Randall has lived [he has lived it as long as Denis and Ingrid]). In exactly the same way that Denis and Ingrid imbibed the memories of *their* grandparents' rooms, through those of their parents, does Randall here imbibe (and express) the memory of a room lived by Nancy and Jasper through the memory of its presence in the room he lives with Denis and Ingrid.

Is any of this related to the injunction of abuse? We cannot be certain. The two parts of Randall's rule *may be* sutured by no more than their common referent, *but* Randall will enjoin "abuse" on only three occasions, twice of plants (22, 43) and here. Is it possible that at age eleven he regarded the stereo—a term (incidentally) never used by Denis or Ingrid—as algesic, as alive? Queried in the present about RULE 100, Randall recalls when "the amplifiers and things used to be kept in the white cabinets."

Uh, never in Denis and Ingrid's memory.

Evidently, though, what is recalled is more than stuff. It is the relationship obtaining between stuff and kid, precisely the substance articulated by the rules: *don't lock your friends out, or your brother* (RULE 40), *close* [the door] *every time you go through* (RULE 4), *be nice* [to the plants] (RULE 86). As we have seen, however, the entirety of these relationships is not caught in the net of the *dos* and *don'ts,* either because the thing is not recognized by the rules (as was true for the sewer pipe [23]), or because the relationship embodies but does not threaten a meaning (rules are emitted only to protect threatened meanings), or because the relationship is embodied in a discontinued (as might involve a potty) or future behavior (as sending mail). Randall's inclusion of such a relationship in his *yes, that's a general rule, but the mail is placed there if you can't put it out* is, as we know, a part of Randall's collaborationist strategy. He will in this way transform Denis's *don't leave anything on the cabinets* into a lesson in Logic *(yes, that's a general rule)* and illustrate it with the exception from Natural History that will prove the rule *(the mail is placed there if you can't put it out)* which, again, deflects the rule from taking Randall as subject while offering up instead the universal *you,* that is, *one,* that is, *everyone.* This shift appears in the coding as a movement from a concern for Randall's (and Chandler's) behavior (kApp) to a description of the role of the object in the operation of the family (tcon), that is, in a shift from the imperative to the indicative, from the normative to the synthetic (Will is disguised as Knowledge). In this way the volitional content of Denis's *don't leave anything on the cabinets* is elided, and thereby *co-opted,* although Denis's commitment to the neat uncluttered whiteness of the cabinet top (only occasionally violated by *his* coffee cup) will be nonetheless *taken up.* Thus Randall's children will doubtless have the opportunity to experience Denis's volition (that is, Jasper's, that is, *Le Corbusier's* "right to be severe toward [the] superfluous") through Randall's reconstruction of it, though the volitive dimension is the one thing that will not be apparent.

Why? Because in adopting appropriate behaviors the kids *do* transform the rules into no more than empirical descriptions of family processes *(that's the way we do things around there):* the volitive is literally . . . evaporated away.

The more readily to precipitate in the world of Randall's children.

So spreads the sway of the neat, the uncluttered, and the white.

XVIII · THE VOICE OF CONVENIENCE

What can it mean for Denis to say *don't leave anything on the cabinets* when in fact that's where the mail goes until it goes out?

Only that here expedience has the upper hand.

Is this generally the case? Evidently not. In fact the rules are emitted precisely to deny the sway of expedience in the life of the family. In the backyard, squirrels nibble nuts and berries all day long, but *we do not eat anytime we want to, we sit down to eat together at mealtimes.* Outside, birds eliminate wherever they are (on the book we are reading), but *we put our wee-wee in the toilet.* It is this above all else that distinguishes human young from those of other animals (that distinguishes human young from themselves as babies), that they distinguish what is advantageous and opportune in the immediate circumstances from what is advantageous and opportune in the long run and under a range of considerations (for example, ethics). Nor is it merely that Nature is invariably expedient where Culture is not (though this is fundamental), but that behavior at home is less expedient than in every engulfing domain. We enter public buildings without concern for the condition of our feet, but *we make sure our feet are clean before coming in the house;* other kids may go to bed when they choose, but *in our house we have regular bedtimes and we keep to them.*

But between the impulsive and the premeditated step many scales of need and desire. Ingrid is *putting away the groceries* and there are things that have to go upstairs. She takes them to the radiator at the foot of the stairs (9) where they will be taken up later under the aegis of another routine (this will be: *going upstairs*). Or Denis has too many books on his desk and a couple need to go upstairs. When the trip is too disruptive of his work, he puts them on the newel post (10): *to take up when he goes to bed*. Or Denis or Ingrid have mail to put out, but it is night or it is Sunday and they don't want it out so long (it is pouring rain, or they are worried about someone stealing it) so they: *put it on the cabinets by the door* (23). Or Denis is reading the newspaper and if he puts his empty coffee cup on the floor beneath the couch he might not see it when he gets up, so he: *puts it on the cabinets by the window* (23). But to let up for even a second is to court disaster: in no time the house is a pigsty, dirty clothes and unwashed dishes are everywhere, the room reverts to the cave Denis and Ingrid first entered whose fireplace (57) was filling up with the stub ends of cigarettes (the prior inhabitants had sat there watching television and smoking, flicking the butts onto the pile on the hearth). But to tighten up is no less problematic: in the middle of writing a sentence, Denis—for a break—takes a book upstairs. There Chandler asks him to *look at this* and would he be interested in a game of Stratego? Coming back down—hours later—he finds the typewriter, still humming . . .

The boundary between the expedient and the deliberative is not fixed, but it is not hard to see that it must be drawn between the coffee cup and the mail: on the continuum between Nature and Culture, the mail on the cabinet constitutes a routinized exception (it is in fact an institution), where the coffee cup remains an exceptional violation (it is the sort of thing Chandler is liable to call Denis on). This boundary has a name: *convenience*—around which cluster acts that justify their commission in its name. Scale is not at issue. Here will be found everything from the taking up of the rugs in the summertime (when the floor [8] is explicitly polished to be a surface suitable for the display of kelim, Navajo, boukara), to the drying of wet clothes on the front radiator (45) in the wintertime (when the rule explicitly forbids putting anything on it), to the slipping beneath the couch (30) of the cup (that is explicitly forbidden to go there). The existence of these acts of convenience—slipping off in one direction toward libertinage, toward precisianism in the other—cannot be denied. Why then do the rules do so?

Because every rule defends a myth.

For example, that Nature will immediately reclaim what is not ceaselessly protected by an active resistance. We already know that this supports an immense labor of dusting, sweeping, swiping, picking up, putting away, tidying up. Here it interdicts coffee cups on the cabinet, shoes beneath the table, glasses on the floor. These it construes as the first signs of an invasion perpetually poised for attack (and indeed, condensate does puddle around the glass, which does leave a ring, which does collect dust, which does destroy the finish, which is, inescapably, the mark of an abandonment that is the beginning of the end of every culture [culture *is* a continuity]). *Don't leave anything on the cabinets* boosts all this into action—defends its importance—especially by ignoring the mail, which it thus forces to declare itself again and again an exception, even in routine, an exception the rule stigmatizes as potentially a violation, the first crack in the dam.

Or it will not be its reclamation by Nature, but its loss to a bourgeoisie. First it's the mail, then there will be a basket for it, sooner or later a doily is sure to appear. Once the floodgates have been opened the room can be wasted by every variety of bric-a-brac, knickknack, gewgaw, gimcrackery, trifle—in a word—superfluity. What then will remain of the claim to the "right to be severe toward those works which serve no purpose, which are superfluous, which are not essential"? In so cluttered a world . . . *nothing matters.* But *don't leave anything on the* [severe white] *cabinets* insures *that it will matter* by repeatedly marginalizing the incursions of the mail. Yes, the mail *is* placed there, but by temporary license necessarily renewed on each occasion (daily, as a matter of fact).

Or it will be simple functional issue: without the rule the cabinets will be so overrun by junk, records not put away, *Tintins,* coffee cups—*whatever!*—that there will be no room for the mail or the mail will not be able to call attention to itself and so will not be put out the next morning (business will suffer).

What is at stake in each of these cases is not only, or even most importantly, the ruined finish of the floor, the doily on the cabinet, the problem with the mail—though decay, clutter, and confusion all *are* resisted—but the myths that catapult these substantial but idiosyncratic realities into signification, into, that is, that larger system of meanings which gives shape not merely to the top of the cabinets but whole realms of family life (and beyond that the social world in general). What the rule defends is the solidification into paradigmatic opposition of Nature and Culture, good taste and bad, chaos and order: the battle between Good and Evil will take place . . . on the top of the cabinets.

It is here that the Voice of Convenience interjects itself, inviting us to admit that between the poles stretches, not a line of division, but a continuous gradation of possibilities along which we may take a step or two, even three or four, without thereby standing on the opposite pole. The mail *can* go on the cabinet and look! Nature has reclaimed nothing (dust is the biggest problem), there is no doily (there is not even a basket), the mail goes out (and if it does not this cannot be attributed to distracting clutter). The room is thereby freed from the responsibility—which structures a formal reception room (a throne room, an apse)—of having *everything* signify. (Everything *does* signify, but the Voice of Convenience permits us to pretend it doesn't.) Without forgoing its seriousness, the room is encouraged to *relax,* is permitted the privilege of *negotiation* between its systemic of meanings and the routines of behavior that evolve not only as Randall and Chandler mature, but as Randall and Chandler and Denis and Ingrid *and the life they live* change. Thus the room is spared a deadly rigidity, a chilling dogmatism, but without in any way impairing the enouncement of the rules which continue to be uttered in an unremitting Absolute.

By permitting interpretation of the rules ("hey, let's not be *silly,* now") (or *"obsessive"*) (or *"hysterical"*), the Voice of Convenience lets us have our cake and eat it too.

XIX · A PLACE IN THE RULES

We have just this minute looked: there is nothing on top of the cabinets but a little Raleigh dust . . . and the monstera. How is it that *don't leave anything on the cabinets* fails to apply to it? Which, as plant, is already a sort of pseudoreclamation, as plant *in* pot *inside* cachepot already a descent into the bourgeoisie, as glad-handing scarecrow inevitably an obscuration of the mail? How come this isn't . . . *swept away?*

Because this is where it goes.

The temptation here is to distinguish between constitutive rules and rules of regulation, that is, between rules that, in a manner of speaking, set the stage—table here, plant there, chair over here (or at least floor here, walls there), and rules that organize the "business"—sit here, speak in turn, *no running in the living room* (or at least no climbing the walls), a distinction, that is, between environment and behavior.

Everything we have said about the room is opposed to this point of view.

Again and again we have argued that the room is *not* a stage set by a designer on which actors recite their lines, but the ongoing expression of a living. We have insisted that this limited liberty of expression is guaranteed by nothing more or less than our willingness to participate in the continuous construction not merely of the stage, but of the theater. In terms of this metaphor—which only gets in the way of our understanding of the room or any other so-called setting—the life of which the room is an expression is *at once* improvisation, acting, play writing, set construction, and theater operation *indissolvably whole* with the environment that simultaneously sustains and is sustained by it. In a phrase: *we live the room.* Nothing escapes the ceaseless change implied by this interpenetration—this co-definition—of behavior and environment. Since there is thus no "set," nothing can be given. Denis and Ingrid entered the room for the first time with a key left them by the realtor. On the windows overlooking the street and neighbor were Venetian blinds hidden by sheers covered by drapes. Valences jutted from walls as an air conditioner from a window. The walls were a bilious institutional green and a dark green carpet covered the floor from wall to wall. In the fireplace a thousand butts slowly decayed. From that moment to the present Denis and Ingrid have not ceased living this room. At no point was it "set"—from which point forward they no more than said their lines—any more than their life was "set" when Denis took his job at State (though there were those who said, "well, you're all set now") or when Denis and Ingrid got married ("now you're set!") or when Ingrid graduated from Wellesley ("you're all *set* now!") or when Denis got his first paper route ("you're all set now," his manager had said as he gave him his route list and delivery bag). This living has been continuous *on all of the scales* which we have seen are implicated in the living of the room, from the diurnal through the broadest stretches of history. Denis and Ingrid have not done so, but the walls could have been moved, the floor chopped (Denis did drill through it), the ceiling opened to the sky (in fact it had been dropped). As it was, Denis and Ingrid had to replace the roof and reconstruct the foundations.

Is there some sense in which these behaviors are constitutive but the taking of an empty coffee cup to the kitchen is regulative? Surely they differ essentially in little more than their frequency, the eternal return of the cup taking its place in the processual dynamics of the day, in turn caught up in those of the week, of the season, of the unraveling years where the replacement of roof and construction of footings themselves appear but a regulative moment in a still larger history of neighborhood recovery and decline. Boylan Heights is currently said to be "coming back," but this also is not given and continuous vigilance must be kept as property changes hands, and biennially resistance must be mounted against efforts by the city to route more traffic through it. What is the difference between this and keeping the cups off the cabinet? Revealing the scale-bound character of the concepts, with each remove the constitutive metamorphoses into the regulative, but in the end, *nothing* is given, life is not Jello, it is not something you eat only when it sets.

What then is the difference between the cup and the monstera?

It is that between living room and kitchen. The easiest way to see this is to ask where if not on the cabinet the cup should be: it should be in the kitchen. Here we see the role of each rule in the geography of the home, which from this perspective is no more than one of *correct locations* (a place for everything and everything in its place). *Don't leave anything on the cabinets* is, after all, hardly unique. But it not only joins the likes of *don't leave stuff on it so people can trip, like marbles* (RULE 48), *don't leave anything on it* (RULE 60), *no leaving things on the landing* (RULE 69), *don't put stuff on top of the speaker* (RULE 97), *don't put things in it* (RULE 105), *don't lay anything on them* (RULE 111), and *don't put things under the couch* (RULE 133)—among others—but works in tandem with an equally numerous group including such positive injunctions as *pick things up you leave on the floor* (RULE 47), *take items on radiator upstairs when going up* (RULE 54), *take up things on the newel post when you're going up* (RULE 59), *Denis and Ingrid hang things on the walls* (RULE 78), *Denis puts the (chooses the) easel items except at Christmas when Ingrid decorates* (RULE 89), *put records back where you got them* (RULE 112), and *put them back when you're done* (RULE 120). On one hand, the *place* (the top of the cabinets, the speaker, the landing, the floor) is protected by a swarm of rules from squatter, emigré, arriviste, or gate-crasher *things* (Lego blocks, coffee cups, mail, marbles). On the other hand, the *things* (records, *Tintins*, easel items) are propelled toward their native places (record cabinets, easel) by parallel constellations of rules. On occasion, as here of the *Tintins* (28), things are both propelled home *(put them back when you're done)* and fended off from foreign shores *(don't leave them out when you're done* [RULE 121]); and now and then places are protected both in general *(as in don't leave stuff on it . . .)* and from particular things *(. . . so people can trip, like marbles)*. The rule pulling the marbles home *(marbles go in the bucket with the marble track)* will not appear in our collection—certainly it will not be associated with the floor—because ours is a collection of living room rules. *Marbles go in the bucket with the marble track* is a bedroom rule (is a rule for the kids' room: it says where *in it* the marbles belong [as the plant belongs on the cabinet]). Similarly the rules for Chassies, Killer, Hundreds, and other marbles games would not appear among the bedroom rules: *you lose your turn if you smooth the dirt* is an outdoors rule, a rule for when you're playing marbles (the only set of rules that is *completely* meaningful is the complete set).

It is for precisely this reason—*that we lack the complete set*—that it appears that plant and cup are subject to different kinds of rules. To enter the kitchen is to realize that the cup too has its place, or more exactly, place*s (dirty cups go on the counter by the sink, only clean Den's cup with the blue pad, clean cups go on the second shelf in the cupboard with the glasses)*, and if the locative plural confounds, recall only that the plant on the cabinet too must be taken out for a bath, must be repotted, must be divided, and that it is therefore caught up in a cycle of activities different from that of the cup essentially in frequency of occurrence. In the complete set of rules (there are literally thousands and thousands of these) every *place* will find its shield against alien attack (as the kitchen cupboards from having the plant put in them) and every *thing* its place (as in *you know where that goes*). In this complete structure rules like *don't leave anything on the cabinets* not only keeps the cabinets pure and white, but keeps everything that doesn't go there . . . up in the air *(that doesn't go there—and it doesn't go there either!! why don't you just put it where it belongs!)*. Identically, but from the other side, a rule like *dirty cups go on the counter by the sink* not only organizes the kitchen for washing dishes, but keeps dirty cups off the top of the cabinets. It is this reciprocity that makes of the thousands of rules a system, one that from this perspective ensures a place for everything and everything its place.

Not, of course, that this place is set. The stereo used to be where the cabinets are, this monstera was once on the other cabinets (26), and before that among the other plants by the speakers (43). *Where it goes* isn't somewhere. It's only a *place* in the rules.

The Cherokee Basket

Bloodroot and butternut stain the river cane Rowena Bradley plaited into the basket glowing in the marigold of the setting sun.

It's a storage basket with a lid, a little more than a foot tall, a little less from side to side. The basket is doubleweave, that is, one river-cane basket was woven inside another in a single continuous operation. This means the pattern inside the basket can differ from that on the outside, and usually it does. In this case, the pattern is the same but the large lozenges are offset inside and out as the pattern slips over the rim without a break. The lid, identically constructed, amounts to another, shallower basket—at five inches, not even all that shallow.

A storage basket. There is nothing inside but air, and a tag by which the U.S. Department of the Interior certifies the Qualla Arts and Crafts Mutual of the Eastern Band of Cherokee as an Indian Enterprise.

rules   103. "Be VERY careful" (Denis, tPRO).
104. "Don't play with it" (Chandler, tCON, tPRO).
105. "Don't put things in it" (Randall, tCON, tPRO).

In the mesh of meanings that structures the room, and toward whose conservation and reproduction the room strains, the basket is a knot. Routes of meaning radiate from it of bewildering variety, among which three are essential: that indicated by the basket as basket, as folk art, and as product of native peoples.

As basket, the doubleweave has a self-evident affinity with the wicker rocker (35) and the plaited fan (59), though had we inventoried the room today, these would have been augmented by a Cherokee honeysuckle basket by Nancy Conseen, an anonymous local oak-splint cotton-picking basket, and a bird's nest Denis found last summer on the ground. At times there have been even more baskets in the room (it is a not unimportant theme), and this varies with need, season, and history. As basket, this product of Cherokee handicraft participates in oppositions between itself and a reservoir of objects that vibrate in the room *through their absence.* It is impossible not to

notice, for instance, that the basket is made of neither a valuable metal nor a precious stone, an observation that through the relay *plant/mineral* energizes *humble/proud.* This is a fundamental opposition at the end of each of the routes we have chosen to follow and it immediately involves the basket in the dialectics initiated by the door-jamb, beveled glass, and quarrels in the sidelights. This container is moreover subsumed, as basket, in the wider theme of the handwoven. This is more dramatically enunciated in the wintertime when the kelim or Navajo stretches out before the door, but even in the summertime it is linked to one of the two throw pillows (36). The opposition here between *handwoven* and *machine woven* (the latter is manifested in slipcase and couch cover [30], wicker rocker cushion cover [35], and the remaining throw pillow [36]) parallels but cannot be reduced to that of *humble/proud,* since handmade remains a prerogative of both upper and lower classes. It is, however, along this route *(handicraft/machine-made)* that the basket joins the wicker rocker in opposition to the David Rowland chair (29), the chrome-plated lamps (34, 42), the electric fan (61), and the glass and stainless steel coffee table (65).

As folk art, the basket enters into conspicuous alliance with the bells (6), the Ferris wheel (25), the Lacandon drum (39), the king (55), the wooden car (60), and the ball game (68), although the bells tend to slip off into folk craft *(art/craft)* and the drum into primitive art *(primitive/folk).* A dominant theme, folk art is sung in the exploitative (the locative paradigm valorizes the objects) . . . but only to assault the pretensions of door, stair, living room, and single-family home (we already know the bell brays in the assaultive). Thus the Ferris wheel commands the large record cabinets (26), sharing the seven-foot shelf with nothing but the basket; drum and king are spotlighted on separate pedestals (38, 54); the car is parked on the hearth (57); and the game is played out on the coffee table (65). What is assaultive about this? It is that these locations—among the most potent in the home, *demanding* mirrors in gilded frames, crystal spheres or obelisks of travertine, bronze statuettes of classical subjects—are occupied by an over-sized toy, a utilitarian basket, a low-fired pottery "drum" lacking every sign of finish. That is, those spaces in the locative paradigm which may above all others be expected to signify *adult* (it is the parent's room [the kids have their own]), *useless* (the living room is *not* the kitchen), and *sophisticated* are spectacularly inhabited by the *childish,* the *useful,* and the *primitive.* Beyond the provision of considerable aesthetic pleasures to those open to them, these objects constitute a two-fisted assault on the values of the bourgeoisie incarnate in the white, two-storied, single-family home: on the one fist, *directly,* through the assertion of value in the *childish,* the *useful,* and the *primitive,* especially here (that is, in the New World) where the "redskin" so vividly embodied the *savage* that the inhabitant of the white, two-storied, single-family home was not; on the other fist, *indirectly,* through an emanant critique of the classical canons of Art (realist mimesis, demonstrable skill, Gombrich's *The Story of Art*) mounted by the ascription to these objects of the appellation "art." Once Art is stripped of its mantle of measurable accomplishment (the "masterpiece") or rarity (the necessarily limited output—however enormous [vide Picasso]—of the individual painter or sculptor), and so shorn of the need for gilded frames, pedestals, velvet ropes, it ceases being capable of certifying measurable accomplishment *on the part of its owner* (taste is revealed as the nominative of an Erotics). This corrosive power of folk art will be fiercely resisted, first by its marginalization (it will not be shown in art museums, but in museums explicitly reserved to folk craft); then by the discovery within folk art of "genuine artists," that is, of veritable masters—that is, named individuals—who invariably will be induced to *sign* their work (which previously had been—precisely—*anonymous*); finally this will limit collectable output and so restore "the art of the people" to its proper owners, the *haute bourgeoisie* and the elite (soon it will be displayed at the Metropolitan). Indicatively, our basket was woven by Rowena Bradley, a named individual, a number of whose baskets—according to the catalog of her first one-person show—"are in private collections and museums."

Nonetheless, the fierceness of this resistance remains a measure of the threat posed by "folk art": what is at stake is nothing less than the definition of "the finer things"—the rewards for measurable achievement (as advertising for luxury goods nauseatingly attests)—where what the room denies through its redefinition of Art is that these are the prerogatives of an elite (turned through a certain angle this would be the strategy of the early Pop artists; turned through another angle, that of the Japanese tea ceremony).

Without denying on the part of Denis and Ingrid the intentionality of this assault, it is essential to stake out its limitations. As was true of the plants (18, 22, 43, 58, 67), this again is a dumb show: the toys are not to play with, nothing is stored in the storage basket *(don't put things in it),* through decontextualization the drum (39) is stripped of its magic. As was true of the Ernst (20), each has become a gem *(be VERY careful, don't play with it),* been transformed into a sign of Denis and Ingrid's *hyper*sophistication, one readily recuperated in the name of individualism: toy, basket, ritual object are from this point of view no more than an eccentricity, one moreover that points to a consumerist *travel to foreign shores* of which these are no more than souvenirs. It suffices to explain to the curious delivery man that Denis and Ingrid have traveled in Mexico: "Oh, neat!"

Satisfaction is complete (it is a trophy room, the folk art are pelts, scalps).

This in no way, of course, mitigates the escharotic power of these objects, though it does serve to demonstrate how effectively they can be recuperated, how readily trivialized. This is searingly evident along the route blazed through the matrices of opposition and affinity by the basket as a product of native peoples. Here the basket is linked in the room as inventoried only to the Lacandon drum (39), but when the other Cherokee basket is in the room, or the Navajo is on the floor, the theme is more compellingly unfolded; and although the walls of the adjacent dining room now are hung with mass-produced (mostly plastic) toys, during the greater part of the time that Denis and Ingrid have lived the home they were covered with Tzotzil and Tzeltal costumes from the highlands of Chiapas: without claiming too much it is fair to insist that the home is alive with a (weak) presence of the American Indian. This presence is experienced, by Denis at any rate, as a slight (but perpetual) doubt about the propriety of his occupation of what is (was, should be) Indian land (that is, about his role in the displacement of the native population), a doubt sufficient to induce enough guilt about his participation in the political life of state and nation (he votes and pays taxes) to constitute a minor (but nagging) anxiety. The basket is especially capable of embodying this guilt (which cannot afford to be overstated, but which should neither be denied), not only because as a product of the Cherokee it immediately brings to mind the bitter betrayal of the United States that led to the Trail of Tears; nor even because it was woven by a descendant of that band that managed to escape the dragnet of Army and militia and hide out in the mountains of North Carolina; but because it so completely incarnates one of the ultimately inadequate strategies pursued in the effort to reconcile native American cultures to those of the Europeans who pushed them aside (that is, who plundered, raped, and murdered them). A century or so ago Cherokee basketmakers were particularly distinguished for their accomplishments in the doubleweave, but—classically—work in the (demanding) art declined until by the 1940s there remained few with the requisite skills or knowledge. Of this time Rowena Bradley has said that when "Mother was teaching me how to weave the doubleweave baskets, there were only two people that I knew of who could do this type of basket: Mother and a very old lady by the name of Toineeta who lived up in the Soco Community." Rowena's "Father would gather the river cane and dig the materials for the dyes for Mother. Everything else was done by Mother" who "tied [the baskets] up in an old sheet and took them to Cherokee to trade for groceries." Through the combined efforts of the Bureau of Indian Affairs and the Indian Arts and Crafts Board working with Qualla Arts and Crafts (the Cherokee crafts cooperative), a revival of interest was encouraged in which Rowena Bradley's mother, Nancy Bradley, was prominent. Unremarked—also classically—is the way in which this revival was achieved by redefining the baskets into objects of art and, as we have seen, shifting their weavers from anonymous craftsmen to showcased artists (not that they were ever anonymous at home). The impression this suppression creates is one of Indians living a life in which these baskets—which we collect and admire for their artistry—remain objects of utility. But at the prices they command (when discussing the rules for the basket, Randall and Chandler both stressed how expensive the basket was) this is impossible, like imagining that the craftsmen who make Lamborghinis or Lagondas could ever afford to own one. That is, the revival was achieved at the price of a certain loss in cultural integrity (the baskets are no longer *used*).

In the room, on the long shelf of a cabinet containing hundreds of vinyl records, the basket is an object simultaneously admired and trivialized. Denied utility, it is a kind of castrato; imbued nonetheless with the scent of an Other, it is sanitized by its pedestalization. Through this presence acknowledged but safely contained, the room wrings its middle-class hands over "the plight of the Indian" while nevertheless taking sustenance from the ineffable quality of its spirit.

And lacerates itself for doing so.

The room: torture chamber.

The Mexican Ferris Wheel

Little kids always ask, "Can it go round?" and in the beginning Denis or Ingrid would turn the crank and run its brilliant colors into a kaleidoscopic rainbow.

It stands four feet high, this *rueda de la fortuna grande* that Denis bought in the summer of 1963 from Enrique de la Lanza Elton. Enrique de la Lanza Elton ran Yalalag—a store in Oaxaca that specializes in folk craft—and he bought it from Pedro Hernandez. Pedro Hernandez made it, from 854 pieces of wood and steel—the latter in the form of the wire from which the seats hang and 607 nails. After he put it together, he painted it in bright aniline colors. The 41 riders get quite a thrill when the wheel is spun, since what with the cabinets the top of the ride is six and a half feet off the floor.

All the cranking has gouged a hole in the end of the axle—and the crank has fallen off—so now it is only in special cases that Denis will reach back to hold the axle on the upright with one hand while he spins the wheel with the other. The colors also have faded, so that what was once a rainbow is now just a blur.

rules   106. "Be VERY careful" (Denis, tPRO).

107. "Don't spin unless you've made sure the wheel won't come off—and it's probably a good idea not to fool around with it at all" (Denis, tCON, tPRO).

108. "Don't play with it" (Randall, tPRO).

109. "Don't turn it, because it's broken" (Chandler, tPRO).

110. "Don't flip benches around" (Randall, kAPP, tPRO).

The Ferris wheel *is* a trophy.

Once it has been ascertained that Denis found it in Mexico, every adult's inevitable question is how he got it home ("it must have been a *huge* crate"). The capture and triumphant return is foregrounded. Lost in this technical discussion is any of the toy.

Evidently it is not a toy (*don't play with it* makes this certain).

There is a smaller Ferris wheel made by Pedro Hernandez and other toy makers of Oaxaca. *La ruedita* stands about a foot high and is turned by a handcrank operating a belt that can drive the wheel forever without eating up the axle. Its existence only emphasizes the extent to which this object on the large record cabinets is not a toy (as the storage basket next to it is not a storage basket). Furthermore this is true in two senses: it is not a plaything for children (it is not a G.I. Joe figure, it is not a Teenage Mutant Ninja Turtle), nor for adults (it is not a fast car, it is not an elaborate stereo system). Yet it is unlikely that Denis bought and kept with him for a quarter of a century a large wooden object that was no more than difficult to bring home.

This exorbitance is nonetheless the essence of the touchstone the Ferris wheel is for the room. This travail, the wheel's size, its location—these all hypervalorize the *idea* of the toy, and hence the *idea* of play. The Ferris wheel is a gigantic and playful symbol—it is patently ridiculous, its harmless excess insists on a smile—that denies the significance of the adult and of work. This is true even for the referent wheels (mimics themselves of the great wheel built by George Washington Ferris for the 1893 World's Colombian Exposition), which absorb energy (you must pay to ride them) but create nothing (there is only the rapidly fading memory of a thrill). In this way Ferris wheels destroy what work creates (wealth). At the same time they arouse a fruitless *fever of excitement* that can only be satisfied by ever-more exhilarating rides: *you spent how much at the Fair?* And indeed the Woods do spend enormous sums at the State Fair where, not coincidentally, they found their Cherokee baskets, the braided rug on the dining room floor, and a taste for clogging and "Orange Blossom Special." At the edge of credibility, Denis will claim that it was only the Fair in the fall of his first year at State that permitted him to imagine it might be possible to live in Raleigh—to believe that there were, after all, countertexts to read beneath the smug pretensions of the North Carolina middle class—and whatever the truth of this every year he does live at the Fair, squandering days among the chicks and rabbits, at the Folk Festival, on the Midway. The Mexican Ferris wheel catches all this up—which is the very populism aflame in *arte popular* (of which the Ferris wheel is the preeminent example in the room), the very populism looking down from Peter's paintings (41, 64), the very populism instinct in the refusal of the lock (7) and the lack of curtains on the windows—and hurls it in the face of the door. If, as we have seen, the door gathers together the threads of pertinent routes through the matrices of opposition and affinity into a single utterance that can only be that of classical patriarchal Western culture supporting and supported by wealth and power, the Ferris wheel similarly draws through itself the threads of an utterance that can only be that of a human utopian populism equally committed to the necessary and the poetic.

Which does the room live? It lives *their conflict.*

The room: battleground.

XX · INNER CONNECTIONS

The room is structured on two levels. The first is largely evident to even the least acculturated among those gaining access to it. On this level we find the prevalent paradigms *front/back, downstairs/upstairs, glossy/matte, fine/crude, clean/dirty.* Let us refer to these as outer connections. When these have been accounted for there remains a residue. This will be experienced as no more than a *je ne sais quoi,* a latency, an irritating discomfort, an indefinable sense of coherence, but at any rate . . . as a flavor. This is attributable to the inner connections.

Here is an example. We have just seen that the syntagm of utopian populism is precipitated in the room by the Mexican Ferris wheel, by, that is, a carnival ride which spins one around through the air. Earlier we saw that insofar as they come through Denis's side of the family the white walls energize an opposition between the *superfluous* and the *essential,* that is, between *wallpaper* and *paintings,* that is, between *decoration* and *dizziness.*

We know that this dizziness, subsequently provoked by the paintings he hung on the walls, was initially experienced in Denis's father by whirling about on the grass. Whirling for the sake of dizziness is thus an inner connection between *l'esprit nouveau* and utopian populism, uniting in the room everything caught up by the Ferris wheel (carnival rides, State Fair, populist paintings, folk art, unlocked doors, bare windows) with everything caught up by the white walls (paintings on the walls, modernist furniture, Modernism, *its* populist instincts), a union moreover which even on this level opposes itself to the door (door, doorframe, beveled glass, bristling newel post) as *dizziness* to *sobriety,* as *carnival* to *Lent,* as *Bacchanalian* to *Apollonian.* Since both sides in this fundamental contest claim the right to be white, the white in the room is animated as a doubt incarnate: the room is tense with an inner ambiguity.

Here's a different example. We have just seen that the Ferris wheel stands in opposition to the doorframe as Bacchus to Apollo. Yet both display the mark of the lozenge, the diamond, the rhombus, the doorframe as quarrels in the sidelights, the Ferris wheel as interdigitated zigzags painted on the uprights. Also netted in this web of association are the Cherokee basket, the wicker rocker, and all the rest of the windows in the room. This is an inner connection. It serves to release from the tyranny of the prevalent paradigms the "room" and its things, to recall the (ultimately) arbitrary nature of the fusion of the signified and the signifier. The inner ambiguity thus induced in the room is liberating: the room relaxes in this inner tergiversation.

A final example. In Peter's painting of Malatesta (41) we find the words "Bow St." These are to indicate the Bow Street Police Court in London where in 1912 Malatesta was tried on a criminal libel charge. In Paul Strand's photograph "Sandwich Man" (56)—seven feet to the east on the same wall—we find the words "30 W. 18th St." Because the open side of the "3" is closed by Strand's framing of the image, however, it is possible to read it as a "B": when this is done the words "Bow St." emerge ("*Bow.* 18th *St.*").

This is the kind of inner connection—first observed by Bob while collecting rules from Randall and Chandler—which André Breton regarded as "a clew between the all too dissociated worlds of waking and sleeping, of outer and inner reality": the room quivers with the unexpected and the unintended.

The Large Record Cabinets

To enter the room is to not even notice the record cabinets. There is the Ferris wheel to galvanize attention, the Cherokee basket, the rest of the room. There are Denis and Ingrid, there are Randall and Chandler. It is not that the presence of the cabinets is unfelt (they are almost seven feet long, more than a couple feet high, and stand out from the wall a foot and a half), but that they have aged with the wall to a uniform shade of pale, that except where the gap opens above the sagging floor, their sides look like records (many of whose spines are white), that they are no more than support, containers, for something other—not, finally, anything themselves.

That's the point.

We, however, already recognize in the twelve Palaset cubes—floating here on genuine Palaset bases—the intrusion into the room of petrochemical feedstocks from the Ural Mountains.

rule   111. "Don't lay anything on them" (Randall, ktAPP).

hand-off   1. "Same as small ones" (Randall, to RULES 100, 101, 102).

Rules are frequently referred from one thing to another. This has the effect, from the perspective of the rules, of subsuming these things in a common, more compre-

hensive class. We have already encountered rules with something of this character. In *plants, don't kill or injure—be nice* (RULE 86) and *just like all other plants—don't abuse it* (RULE 99), the peace lily in front of the easel and the monstera on the small record cabinets are subsumed in the more general category *plants*. In *same as all pictures— don't put your hands on it (same as windows)* (RULE 90) this happens twice. First the Ernst is subsumed in the class *pictures and windows* (not *glass* because the rule for the Mies van der Rohe table is the comparatively weak *try to keep your fingers off the table* [RULE 211]). Furthermore, this is implicit whenever a rule is repeated, as when *be VERY careful* is applied identically to the Cherokee basket (RULE 103) and the Ferris wheel (RULE 106): here the inclusive class is *fragile art object*.

It is here, then, with the large record cabinets, that we come across the first instance of an explicit *hand-off* of the rules of one object to another: *same as small ones*, Randall says, thereby applying to the large record cabinets the rules previously educed for the small (that is, RULES 100, 101, 102). The appearance of these HAND-OFFS in our text is an artifact of our collection of the rules. Were Randall to leave something on the cabinets, Denis would not be likely to say, "Pick that up. You know the rules. They're the same as for the small ones," but "Pick that up. You know you don't leave anything on the cabinets!" emitting the rule . . . as such. Indeed, Denis gave precisely this rule *(don't leave anything on the cabinets)* here, making implicitly the connection that for concision (and to assert his collaborationist's mastery of the rules) Randall makes explicit, namely, that as members of a common class *(record cabinets)*, the cabinets are subject to common rules (100, 101, 102). We therefore did not assign a rule number to Denis's *don't leave anything on the cabinets*, since this is perfectly caught up in the hand-off (it is RULE 101), but did to Randall's *don't lay anything on them* to respect the difference between it and RULE 101 (that is, the difference between *lay* and *leave*). This might not be real: Randall may have added it not by way of supplementation, but illustration.

That the hand-off was from the small cabinets to the large is similarly an artifact of the arbitrary order of our inventory. Historically the rules originated for the large record cabinets which Denis and Ingrid brought with them when they moved into the house. Nine years later the small record cabinets entered the room, attracting to themselves the rules that had been evolved for the large ones. Because, however, the new ones ended up near the door, they also attracted to themselves the historic function of the "stereo" cabinet to be a place for the mail when it can't be put out. Because mail is *not* placed on the large record cabinets when it can't go out, Randall's *same as small ones* is not quite right.

What goes on the cabinets is the Ferris wheel, and the Cherokee basket.

THE BEATLES LET IT BE
THE BEATLES · HEY JUDE
LONELY HEARTS CLUB BAND · THE BEATLES
ENDLESS SUMMER
BEACH BOYS / PET SOUNDS / CARL AND THE PASSIONS – SO TOUGH
THE BEACH BOYS L.A. (LIGHT ALBUM)
SPIRITS REJOICE
ALBERT AYLER IN GREENWICH VILLAGE
A-9155
THREE EVENINGS WITH FRED
RECORDED FROM THE SOUND
FRED & ADELE ASTAIRE
MR. TOP HAT
FUNNY FACE

The Records

Sometimes when you enter it, the room is filled with music. In the summertime you usually know this beforehand—the music can be heard on the sidewalk, it can be heard on the street—but even then to listen to it is to attend to this sound instead of that. The birds in the trees do not stop singing because Mozart is on the turntable, the fan in the kitchen window hums in and out of awareness just the same.

When a dump truck passes, it is memory alone that can complete the passage.

None of this is independent of the room. The room may not require the birds to sing, but its open windows mean they must be heard (the room is not air conditioned). The dump trucks do not exist to kill the music, but the room is not accidentally on a busy street (it is downtown). The Mozart may not be the only thing heard in the room, but the possibilities are not limitless: the room has "a sound" and if the birds and trucks are two among many components, the most characteristic stems from the records stored in the record cabinets.

There are 1,720 of these. When the room was inventoried, there were fewer; there were 27 *Tintins* (28) in the lower right cube of the smaller cabinets (23), but these have been moved upstairs, and now it is all records from one end to the other. They are arranged like this. In the first cube—upper left-hand, large cabinets (26)—there are 66 Christmas records. The rest of this cube and the next ten—counting across and then down—contain "classical" music from the Western tradition, arranged alphabetically by composer, and within composer either chronologically (as Stravinsky) or by form (as Charles Ives). The next seven cubes—it starts again top left with the smaller cabinets—contain twentieth-century popular music and jazz, arranged alphabetically by performer (as Blossom Dearie), composer (as Harold Rome), title (as *Show Boat*), or genre (as "old gold") depending on how it is referred to. The Woods say, "Let's listen to *Pajama Game*," not "I'm in the mood for Adler and Ross" (so *Pajama Game* is filed by title). The last two cubes contain folk music arranged alphabetically by country (India) or geographic region (the Caribbean). There is some "classical" music mixed in here which the Woods refer to by country of origin (for example, Manuel Ponce is filed under Mexico [though Chávez and Revueltas are in the "classical" cabinets]) and a lot of the "folk" music is really popular (Oscar D'Leon is filed under Venezuela).

Despite the powerful focus on European and American classical music of the past three hundred years, music can be found here from sixty-two countries (not distinguishing musical subcultures, as Hawaii and Puerto Rico in the United States) and ten centuries (ignoring an attempted reconstruction of the music of ancient Greece). Though not large as record collections go, it nonetheless weighs half a ton. Though half of this is packaging (paper, ink, cardboard, assorted plastics), the balance is polyvinyl chloride, six billion pounds of which are annually produced in the United States alone from the carcinogenic gas vinyl chloride. Vinyl chloride in turn is derived from the petrochemical feedstock ethylene. Given recordings from France and England, Mexico and the United States, Venezuela and Germany, Holland and Brazil, it can be taken for granted that the room dips its wick in the oil fields of every major petroleum-exporting country in the world.

This is the infectious sound that must be heard in the room: the beat of the walking-beam pumping crude from the ground, the hiss of the gas . . .

rules   112. "Put records back where you got them" (Denis, tCON).
113. "Put them away after use—" (Chandler, kAPP, tCON).
114. "Unless you don't know where they go" (Randall, tCON).
115. "Don't leave them around" (Denis, kAPP, tPRO).
116. "Make sure you put sleeves on records so that opening is up" (Randall, tCON, tPRO).
117. "Be careful not to scratch" (Randall, tPRO).
118. "Be careful in general" (Chandler, ktCON, tPRO).
119. "Don't use them as Frisbees" (Randall, kAPP).

What is a quarter ton of vinyl chloride doing in the room?

Taking the place of the piano. Actually, the piano is in the dining room. Still, it is rarely played, and the point is that in general the Woods play records, not instruments. That is, they listen to music played by others. Furthermore, this experience is radically mediated. Seen from this perspective, the records represent a quarter of a ton of concentrated alienation.

This is burned away in the heat of a consumerist euphoria: not large, the collection will be . . . *good.* The praxis of the Woods is to choose (and [incidentally] to buy). The record collection is a consequence of choices. Not independent of each other, these comprise a *system of choices,* that is, they embody the Woods' *taste.* That this is true of the room as a whole and in all its parts goes without saying, but because the record collection has few of the alibis that recuperate so many of the things in the room (the records are not useful like the wicker rocker [35], they are not heirlooms like the easel [19]), and because of the profusion of distinctions encouraged by the divisions into genres, regions, periods, styles, composers, performers, the record collection forces taste . . . *to disrobe.*

This taste is not innocent (the Woods have spent more money on the record collection that exhibits it than on everything else in the room combined). Yet this innocence is precious—it is the final refuge of the Natural ("I [just] like it")—so it is heavily defended. The variety of these defenses—"I know what I like and I like what I know," "It's just not Good art!," "I hang paintings on the walls about me to help create that dizziness"—does nothing to obscure their class basis, that is, in available economic and educational capital, a basis moreover that supports the objects selected in this innocence no less than the defenses erected to protect it. Without attempting a taxonomy of the many class fractions characteristic of contemporary American society, we may nonetheless say that the taste exhibited by the record collection is . . . *high.* First, because it is a mark of *ease,* that mark exploited by the bourgeois above all others to distinguish the *disinterested* from the *interested.* This is an opposition fundamental to a family of oppositions, some of which we have seen (most obviously *living room/kitchen,* that is, *mind/hand* or *soul/stomach* [unavoidably, as we know, within the home, *Denis/Ingrid*]); but others of which we have not, or not seen in quite this way: *place of residence/place of work* (in this context indistinguishable from *living room/kitchen*), *holidays/work days* (an opposition filled with animus in Peter's hatred of the Sundays on which Jasper would listen to "that music" [Schoenberg]) and so *leisure/work* (that is, *living room/kitchen*) to say nothing of other classically Weberian distinctions including *art/industry, sentiment/business,* and *world of artistic freedom/world of economic necessity,* where we find ease—that is, the record collection—on the side of *artistic freedom–art–sentiment–leisure–holidays–place of residence–living room–mind–soul–Denis.* Here it is crucial to acknowledge that Denis chooses which records to buy (in general) and play (most of the time). The paradigms are *not* ridiculous abstractions or irrelevant reifications, but the structures of intensely lived and vividly real behaviors.

The disinterested quality that is the hallmark of real ease is doubly marked (this is the second sign of the high taste exhibited by the records). The works that dominate the collection are not merely uncommon—that is, not to be found on the tape deck in the car of every Tom, Dick, and Harry (for instance, Gunther Schuller's "Conversation" played by the Modern Jazz Quartet and the Beaux Arts String Quartet)—but *abstract* (Bach's *The Well-Tempered Clavier*), *austere* (the *Six Bagatelles* of Anton Webern), *demanding* (Albert Ayler playing his "Spirits Rejoice"), or all three (Peking opera). Above all, they have been canonically enshrined: Bach's *Mass in B Minor,* the *Brandenburg Concertos,* Beethoven's late quartets, his symphonies (but as performed by a chamber orchestra), Handel's *Water Music,* his organ concertos, Haydn's trios, Monteverdi's *Marian Vespers* of 1610, *The Magic Flute, The Marriage of Figaro, Così fan tutte,* Schubert's songs, *Le Sacre du Printemps.* The record collection in large part can be characterized as one of consecrated refinement.

The third sign of high taste is the freedom the collection claims from being bound by its own taste; that is, its being in the position to refuse any and all refusals. Here the amply exhibited "good" taste is exploited to "legitimate" preferences for works beyond the pale, that is, for the *common* (Michael Jackson's "I Saw Mommy Kissing Santa Claus," his *Off the Wall,* his *Thriller*), the *programmatic* (Beethoven's *Pastoral* symphony, Charles Ives's *Three Places in New England*), the *lush,* the *exuberant,* the *warm* (a Charpentier *Te Deum,* Schoenberg's *Gurre-Lieder,* Bizet), the *undemanding (Peter and the Wolf),* or all combined *(Bolero).* Above all, these will have been aggressively proscribed by the arbiters of good taste (Denis owns all six of Shaun Cassidy's records, one of Leif Garrett's, a copy of Merle Evans's *Circus in Town,* Elmer Bernstein's soundtrack for *The Sons of Katie Elder* ["including 'Texas Is a Woman' recited by John Wayne"]). This is a demonstration of the populist "good faith" otherwise in evidence in the portraits of revolutionaries (41, 64), the Ferris wheel (25), the unlocked door (7) (in none of this is there a trace of parody, of cynicism). Reaping the pleasures of good taste while ignoring its obligations is always evasive, a posture which as we know will oblige the collection to saturate as many paradigms as possible. While the extent to which it does so cannot be exhaustively presented (the collection has twenty-five times as many terms as the rest of the room), it will be necessary to point out a number of the more salient saturations.

We already know, though it does so in varying degree, that the collection saturates the paradigm *folk* (10 percent of the collection)/*popular* (35 percent of the collection)/*elite* (55 percent of the collection). Oppositions within each of these terms are equally saturated. There are 50 records of medieval and renaissance music; 54 devoted to the music of Bach; another 54 to that of Mozart; 41 to that of Beethoven; and 54 (again!) to that of Stravinsky—alluding only to those categories with more than 20 records in them—thereby saturating the paradigm *medieval/renaissance/ baroque/ classical/romantic/modern.* Within the latter we can distinguish—among others— *tonal/atonal* (*Weill* [21 records]/*Schoenberg* [25 records]), *through composed/aleatoric* (*Britten* [20 records]/*Cage* [20 records]), *European/American* (*Satie* [20 records]/*Ives* [52 records]). The affinities along one side of these oppositions (atonal-aleatoric-American) begin to sketch a syntagm, that of the American experimental tradition: Charles Ives, Carl Ruggles, Henry Cowell, Harry Partch, Edgar Varese, Morton Feldman, Lou Harrison, John Cage are massively represented, whereas their more academic counterparts (Bernstein, Copland, Sessions, Babbit, Block) are not, or only by happenstance (they are on the flip-side). This acknowledges a paradigm *academic/ nonacademic* which the record collection *does not* saturate, allowing it to come down emphatically on the side of the street. This is lived out in the 35 percent of the collection devoted to popular music, within which such oppositions as *American popular song/rock and roll* and *Motown/California sound/British invasion* are deeply impregnated (*Fred Astaire* [27 records]/*"old gold" rock* [45 records]; *Motown* [20 records]/*Beach Boys* [21 records]/*Beatles* [22 records]). With at least sixty ethnic groups represented in the folk collection, terms in any number of paradigms are drenched, soaked, and steeped. What unifies a congeries such as this is not some "essentialist" quality (though when Denis is asked what kind of music he likes, he does say "*good* music"), but: the *freedom* the elite has always claimed as its most definitive trait (it is disinterested to the point of abandonment). Is this to deny an aesthetic universalism to Denis and Ingrid? Not at all. It is merely to note that such a universalism is as much a social construct and class attribute as any parochialism.

The key to the conservation and transmission of the taste exhibited through the collection are the injunctions to *put the records back where you got them,* to *put them away after use,* and *unless you don't know where they go.* In contradistinction to *don't leave them around,* which we understand to be opposed to the return to Nature promised by every form of sloppiness, and the injunctions *to care* needed to preserve the physical collection itself, these rules confront the kids with the collection as taxonomy, one which constitutes not merely a mathesis (with its implied curriculum) but an appropriation. The collection after all is a kind of encyclopedia. The music is never heard solely in itself, but as representative (as well) of region, genre, period, style, composer, performer. Any cover will elicit the question, "Do you know who played this originally?" The Clash's "Spanish Bombs" will provoke a lecture about the Spanish Civil War. At dinner Ingrid rhetorically will ask, "What are we listening to tonight?" Presented in 1,720 elaborately ordered, colorful spines, the lesson is that the world can be mastered, or at least the world of music (but it *is* a synecdoche): here it is, in these cabinets in the living room. For the kids to master it, *all they have to do is learn where the records go.*

Is this a self-conscious effort to raise kids with culture? Not particularly. It is a by-the-way adjunct to the selecting, pruning, playing, sharing, enjoying, dancing to, singing along with that is the collection in life, one that without doubt is the privilege of kids growing up in a household at a certain level of educational and economic capital.

Is any of this to deny that the music is beautiful? No. It is only to acknowledge that beauty too is a social construct.

XXI · INTEGRATION

We know that it is sufficient to maintain the doorframe (3) in the best possible condition to act out all the implications of its form. This is because the measured and temperate behavior demanded by the rules embodies the Apollonian character embedded in the door. The same is true for the records (27). It suffices to put the records back where they belong to live out the implications of the collection's taxonomy, not merely its division into elite, popular, and folk (the record collection as model of class structure)—and so forth—but the appropriation implicit in no more than the fragmentation of the world of music into finite objects named and classified (that is, subdued).

This is not the only way the collection will be integrated into the lives of the kids. Recently, for example, Chandler asked Den to play *Where In the World Are You?* with him. In this, Chandler has to guess the names of countries blanked out on maps in a deck of flash cards. Randall joined in mid-game, trying to guess the places Chandler was unable to. Denis was vocably disturbed by the lack of knowledge demonstrated; which led to a discussion of kinds of knowledge he had had at the kids' ages (stories of walking home with Ashley Abramson identifying countries in the outlines of puddles, patches of dirt). Denis also told them about a curriculum he had just read about involving the drawing of world maps from memory. The next day Randall came home from school with parts of a world map he had drawn in his free time. Denis had spent the morning trying to determine the number of different countries represented in the record collection (if it hadn't been this it would have been something else). The two trajectories intersected. To help Denis, Randall ended up massaging an atlas; and that night instead of The Clash or Big Audio Dynamite he listened to Tibetan tantras while he did the dishes. (The effect lingered: the next night he listened to Irish pipe music). The taxonomy controlled by *put the records back where you got them* here functions as a relay between the auditory experience of the music and geographic knowledge of the world: field of rules, child, world . . . *fuse*. What is analytically articulated is, in the experience within, *whole*.

A third way: for the kids to see it live in their parents. Denis and Ingrid have friends from their dance class over. The rugs are rolled up, the couch (30) pushed out of the way. Records and room enter into a tender relationship, as if it were they who were dancing.

The room: music hall.

The *Tintins*

Mostly the spines are red and yellow. Three are green. Many are cracked, broken. A couple have come apart at the top. All are worn, all have the look of having been much used.

It is unusual to find them all on the shelf. Usually one, often two, are lying around,

on the couch (30), in the seat of the rocker (35), on the table (65), on top of the record cabinet by the door (23), when, that is, there is not a kid or two and sometimes Denis reading them, *sometimes* Denis only because he has read them all so many times he no longer has to read them often, not more than once a year anyhow.

Strictly speaking, they are not all *Tintins,* not even all Hergés. There is, for instance, the massive *Le Monde d'Hergé,* which is by Benoit Peeters, and the issue of *The Comics Journal* with the long article on Hergé by T. F. Mills. There is also *Le Musée imaginaire de Tintin,* which is the catalogue of a show held in 1979 at the Palais des Beaux-Arts in Brussels, and one of the adventures of Jo, Zette, and Jocko *(Le Testament de M. Pump).* Otherwise all the *Tintins* are here that have been published in book form since *Tintin au pays des Soviets,* twelve of them in English—six in the wonderful Golden Press edition, five from Methuen, one of the paperbacks from Atlantic–Little Brown—and ten in French—from Casterman. There is even an *album du film,* Casterman's *Tintin et le lac aux requins.*

rules   120. "Put them back when you're done" (Randall, ktcon).
121. "Don't leave them out when you're done" (Chandler, kapp).
122. "Don't pull them out by the top of the spine—" (Denis, tpro)
   "—right, by the middle of spine" (Randall, tpro).
123. "Treat with care" (Denis, ktcon, tpro).

Denis's version of RULE 120 is a little more comprehensive: *put them back when you're done reading them or—I'll sequester them.* Why such force? Because they were always being left out, the kids would be reading them, the doorbell would ring, it would be a friend, they would rush off leaving . . . *Tintins* everywhere. Or the friend would be reading *Tintins* with them, Ingrid would ring the bell for dinner, and *Tintins* would be left . . . everywhere. Or it would be time to close up downstairs. Denis would cut out the light in the kitchen, fluff the pillows on the couch (30), and . . . *put away the Tintins.* That is, for Denis this situation presents a twofold transgression. The first we know: it is that except for ceaseless vigilance the barbarians would have been over the wall yesterday. This is reflected in the usual assignment to the appearance code of Chandler's *don't leave them out when you're done* (civilization is putting things away [barbarians just leave them around . . . anywhere]). The second is new; it is that his kids *are* barbarians and he is forced to pick up after them. This is reflected not only in the assignment to both sides of the control code of Randall's *put them back when you're done* (civilized kids are those who put things back), but in the threat of Denis's version: *or I'll sequester them.* Inasmuch as sequestering the books *would* relieve Denis and Ingrid of having to pick them up (or, more often, call the kids back to put them away themselves), but *would not* advance the march toward civilization on the part of the kids, it marks a kind of paralysis, one likely induced by the frustration of a certain stage in the development of the kids, that in which it has become evident that the kids are not barbarians (any longer [for trailing clouds of glory or not they come as such])—*and so do not have to be picked up after as a matter of course*—but in which they have *not yet fully mastered* the refined arts of a cultivated life—*when as a matter of course they will be able to pick up after themselves.* This stage is of varying length (it is never short) and is passed in different domains at different rates and times. That the kids *do* pick up after themselves *here* should mean that they *could* pick up after themselves *there,* except that this is the logic of philosophers (it is adult logic) not that of development; and it is *this* tension, between the logic of maturation and socialization and that of the books and schools (or at any rate, operational logic) that results in the frustration giving rise to the threat. From this perspective it is seen as springing from the gap between the kids' attainments and Denis's aroused expectations.

Here these expectations happen to be fueled by the very objects of the rules (the subjects are always the kids)—that is, *books,* especially in a home where these are not merely the mark of civilization they may be elsewhere in the world but the objects of an elaborate fetish. Note that the first of the Golden Press *Tintins* were given to Denis and his brothers for Christmas in 1959, that Denis has clung to these physical artifacts for thirty years, that during the 1960s the acquisitions of further *Tintins* involved him first with British and then French bookstores (problems of translation, currency exchange, customs), that he has been reading and rereading them ever since. Of course, the kids weren't, or certainly not in the same way. It is the glory of the *Tintins* to be able to be read at many levels. On the first are the utterly absorbing illustrations (most in evidence in the magesterial *Destination Moon*). Kids who have sucked these dry move on to the visual humor streaming across increasingly large numbers of panels (such as the "sticking plaster" joke that constitutes a subtext for whole pages of narrative in *Vol 714 pour Sidney*). Next they learn to appreciate the verbal *faux pas* of Thompson and Thomson (or Dupont and Dupond) and laugh themselves sick over each next encounter with "Cutts the butcher." Somewhat later they begin to grasp episodes of plot (the exploding gasoline, the mirages, the hair growth in *Tintin au pays de l'or noir*). And later still the plots of the books as wholes begin to clarify. Hergé's larger project, its relationship to European colonialism, the real brilliance of the drawing, these are perhaps levels still beyond Randall and Chandler, but at any time the level on which they were reading was never fixed, it was always one level up, two back, so that what was covered by "they're in the living room reading *Tintins*" was at best an open description of what was in their hands and the posture of their bodies in the couch (30) and the chair (35).

What were the books doing in the living room? They were the kids' hold there, they were *for* the kids, something in the room *for them,* that was theirs. When things got tense in their bedroom, or it was time for one or another of them to cool off, Denis or Ingrid could always say, "Sweetheart, why don't you go get a *Tintin*?" and soon enough the kid in question would have vanished through the rabbit hole of Hergé's imagination that yawned in the living room. To drop back a level or two is like . . . *going home,* a load is lifted, the order of the world restored. Just this afternoon Randall—close now to fifteen—was discovered in Denis's study laughing out loud over the "sensational entrance of the Thompson twins" in *Destination Moon;* nor was it many nights ago that Chandler implored Denis to answer the phone, "Cutts the butcher!"

As for Thompson and Thomson's "to be precise" and "I might even say"—these have become stock phrases in the everyday speech of the kids.

XXII · THE BARBARIANS

What is a barbarian?

He is uncivilized. Why is this? *He does not speak.* Etymologically this is quite clear. Our word comes from one the Greeks applied to foreigners to express the strange sound of the language which was . . . *not Greek.* What was this sound? *Baby's babble.*

Etymologically this also is clear. *Baba-,* the Indo-European root, is a mimicry of unarticulated sounds, indistinct speech. In time it exfoliates into babble and barbarian, but also into booby (from the Latin *balbus,* stuttering, stammering); baboon (from the French *babine* and *baboue,* both associated with incoherent speech through the pendulous lip, the grimace); babe, baby; bambino (from the Italian *bambo,* child, simpleton); baba, babka (from the Polish *baba,* old woman); babushka (from the Russian *baba,* old woman).

What is defined by this field of associations?

The Outsider, the Other: *he who cannot speak* (he babbles, he stutters, he stammers, his jaw is slack, he makes faces, he drools [he is a baby, an idiot, he is an old woman, an animal, he is a foreigner {he is the other}]).

Randall and Chandler are Outsiders (they are barbarians [they are babies]). It is the work of Denis and Ingrid to bring them in, to erase the signs of the Other, to raise them from the animal, to teach them to speak: *to teach them the rules.*

To give them Culture.

The David Rowland Chair

Foam, foam of air and light, steel and flame. "Can you sit on it?" people ask. "Is it really a chair?"

In a tubular-steel frame David Rowland has mounted a seat and back of steel-wire furniture springs coated with a bright red elastomer of polyvinyl chloride. This "fattens" the springs, keeps them aligned without making them rigid, lets them stretch slightly when subjected to weight or pressure. It also makes them "warmer" to the touch. First manufactured by Thonet in 1979 (in their Statesville, North Carolina plant), the Sof-Tech chair is not only the most up-to-date piece of furniture in the room, it is the most modern.

"Sure it's a chair," says Ingrid. "Try it."

rules   124. "Treat it like a chair, not like a toy" (Denis, kapp, ktcon).
125. "Don't bounce on it" (Chandler, kapp, ktcon, tpro).
126. "You fall more easily than on a normal chair if you stand on it" (Randall, kpro).
127. "Put it back when you're done" (Chandler, ktcon).
128. "If you bicker over who'll sit on it you can't use it" (Ingrid, kapp).

What is the chair that the kids would bicker about it? It is a toy.

This is the meaning of the steel-wire furniture spring, that the chair is a hobbyhorse, that it is a Slinky, that it is a pogo stick (the room is a rumpus room, it is a playpen). It thus denies everything insisted upon by the window in the door (4), the doorframe (3), the glass in the sidelights (5). The chair is not solid (it has no burgher pretensions). It is a bright red (it is provocative, it is not somber, it is not self-serious). It is metal (it is not wooden, it is not upholstered, it is not traditional). Is this serendipitous? Here is a collaborator of Le Corbusier: "Metal plays the same part in furniture as cement has done in architecture. It is a revolution. . . . Brightness—loyalty—liberty in thinking and acting. We must keep morally and physically fit. Bad luck for those who do not." But here is David Rowland himself: "Steel wire fits a concept which I believe to be fundamental to good design—to accomplish the most with the least." It is as with the white walls: the restraint, the efficiency (the elegance) are classicizing; the revolution implicit in the exposed metal is romantic: the bourgeoisie is again zapped from both sides (it will arise again . . . *with modern furniture*).

The room beyond the door will not be the one promised in the door; but neither will it be the one called for by the chair (by the white walls). The room will live their confrontation, their confusion, their contradiction.

The room: dialectic.

The White Couch

"PH-E-E-E-T"

"What's that siren for? An engine on fire! That's the alarm for the extinguishers!"

"Thundering typhoons! The extinguishers haven't worked; it's burning more fiercely than ever!"

"Wadesdah tower . . . Wadesdah Tower . . . This is KH-OZD . . . Starboard motor on fire . . . Extinguishers unserviceable. We're turning back . . . We'll try to reach Wadesdah."

"TICK TOCK TICK TOCK—"

But wait, this sound is not coming from the luggage compartment where Snowy is trying to get Tintin to find the bomb in the doomed Arabair flight to Beruit but . . . *the porch?* Chandler's eyes leave the pages of *The Red Sea Sharks* to look over the back of the couch toward the door. There *is* someone there. Sitting up he throws his feet to the floor. As he pushes up from the seat cushion with one hand, his other keeps his place in the book. Entering the room from the couch, he pads toward the door.

The couch is all white, except for four wooden legs so unobtrusive against the floor that the couch seems to levitate. You wonder how they can hold it up, the couch and the kids and the adults, but they do. It's not a huge couch, but it is six and three-quarters feet long: four people can sit on it at need, three to unwrap presents, two to talk, one to stretch out and read *Tintins*. It's two and three-quarters feet deep and two and a quarter feet high in the back, with an armrest twenty-one inches off the floor. It's not a low, *low* couch, but there's no chance of sliding off it either. If the insides retain the authentic baroque of cambric, springs, stiles, rails, decking, edge rolls, batting, burlap, stuffing, arm posts, arm boards, arm braces, corner blocks, welt cord, blind tack tape, webbing tacks, nails, and twine that has characterized upholstery for a hundred and fifty years (and more), the outside presents a clean simplicity no less characteristic of Paul McCobb, who designed it in the early 1950s. This simplicity is especially evident in the white slipcase Ingrid made for it, despite the three back cushions, three seat cushions, and case for the couch body proper, each with its snaps, 20 across the backs, 24 in back of the seats, 118 along the bottom. There *are* fewer parts to the couch than there are records in the collection, but not many.

"*TICK TOCK TICK TOCK.*" This time it *is* the bomb in the luggage compartment primed to blow KH-OZD to smithereens. Back on the couch, Chandler has rejoined Tintin in the Middle East and . . . *slipped from the room.*

rules   129. "Don't throw yourself onto the couch—sit down on it" (Denis, kAPP, ktCON, tPRO).

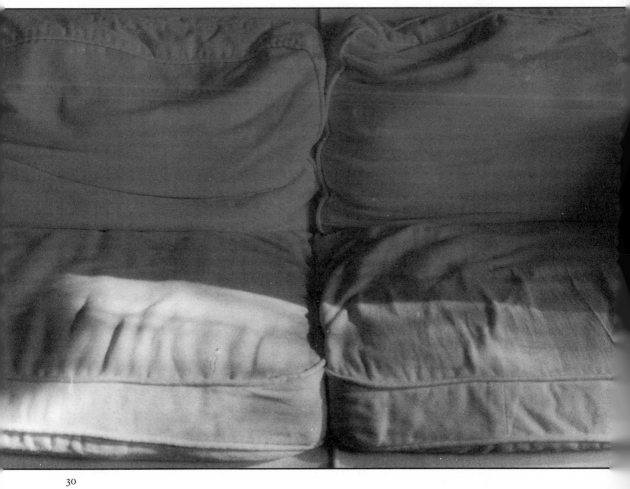

130. "Don't sit on the arms" (Chandler, Randall, kAPP, tPRO).
131. "Don't sit on the back" (Chandler, kAPP, tPRO).
132. "Don't hang off back of couch" (Denis, ktAPP, tPRO).
133. "Don't mark back with shoes" (Chandler, ktAPP, tPRO).
134. "Don't climb over the couch" (Chandler, ktAPP, tPRO).
135. "No fighting on the couch" (Denis, ktAPP).
136. "No shoes on the couch" (Chandler, ktAPP, tPRO).
137. "Don't put food on it" (Chandler, ktAPP, tPRO).
138. "Don't throw up on it—go outdoors or in the bathroom" (Randall, tAPP, kCON).
139. "Don't get the couch dirty" (Denis, tAPP).
140. "Don't hide things under or behind the pillows" (Randall, kCON).
141. "Don't take up the pillows and sit underneath them" (Chandler, Randall, kAPP, tCON).
142. "Don't leave the pillows not spiffed up" (Randall, tAPP, kCON).
143. "Don't leave things under the couch" (Chandler, ktAPP).
144. "Don't leave your shoes under the couch" (Denis, kAPP, tCON).

The couch is the site of another amphibology: constructed as a chair, it is stretched into a bed. Any attraction it might have is doubled by this doubling of its potential interpretations: it is a bed in a chair, it is a chair-bed, it is for sitting up–lying down, it is for throwing feet over the back, it is for climbing on, it is a jungle gym, it is a playground, it is a grappling place, it is a place for holding, for making out, it is a place for making love, for being discovered, for sitting down and having serious talks, it is a place for scenes. It is a twilight place, neither public nor private. With the stair treads (12) it undoes the distinction elaborated by splitting behavior . . . between two floors.

This particular couch has subverted these and other paradigms in five rooms. Denis's parents bought it in 1957 when they moved from a public housing project on the Cuyahoga in downtown Cleveland to an apartment in Cleveland Heights on a street overlooking the city. It replaced an old studio couch: it was a big deal. It wasn't only that it was upholstered, that it had a back *and* seat and back cushions; or even that it was designed by Paul McCobb, at the time an arbiter of good taste (he was being regularly shown in the "Good Design" exhibits at the Museum of Modern Art [Bloomingdale's had a Paul McCobb Shop]); but that as such it incarnated the Woods' transition from "bohemia" to the burbs (*studio couch/couch* :: *housing projects/ suburban apartment*). That such a transformation should be embodied in a couch is no more than expected: were it new, even in its current situation it would be the most expensive piece of furniture in the room; and furniture has been, from the beginning, among the clearest signs of wealth and power (etymology: *divan,* among the Otto-mans the seat used by an administrator when holding audience).

When the Woods moved to a double house they took the couch with them, eventually reupholstering it in a loose-woven rough pinky-yellow fabric Denis's par-ents found in Mexico. They did not, however, take it with them when they "retired" to Mexico (they did take the Saarinen "Womb" chair and ottoman, the single-pedestal Saarinen coffee table), selling it to Denis and Ingrid, who took it with them to their apartment in Worcester. Although it thus entered its third room from Denis's side of the family, it could have entered from Ingrid's, for the Hansens too had bought a Paul McCobb couch (theirs was from his "Directional" line), also on the occasion of a move, from a house in Greens Farms (Connecticut) to an apartment in Greenwich. We find ourselves here in the presence of a third convergence between Denis's and Ingrid's sides of the family (that of Mexico was the first, that of the white walls [16] the second), one which suggests that whatever the difference between the Hansens' neo-Biedermieresque German-American Art Deco and the Woods' more doctrinaire but no less imported Franco-German Modernism, it amounted to little when it came to signifying similar positions on parallel trajectories through the class structure of American society.

By the time Denis and Ingrid acquired it, the upholstery was already worn and soon enough Ingrid felt compelled to make arm covers out of remnants of the Mexican material. On a subsequent trip to New York, Ingrid and Denis were taken with (but what can this mean?) a white couch they encountered in a Soho gallery peddling Guatemalan textiles. Because of her strong feelings that it would be more convenient to clean a slipcover than upholstery, Ingrid opted for the former (resounding echoes of her mother), no doubt a peculiarly relevant consideration given the characteristic of white to be that color above all others suited to receive the mark of the dirty, and so obligated to participate in the economy of the clean (here the couch allies itself with the floor [8] and the glass [4, 5, 17, 47, 50, 65]). This whiteness enhances the ability of the couch to play the central role ordained for it by bourgeois principles of interior decoration, since—as we have seen—nothing is more capable of signifying the Cul-tural (in this instance: superb home care skills) than *the absence of dirt,* the more readily displayed on a surface uniquely prepared to draw attention to its presence.

It was this sign of haute-bourgeoisie pretension that Denis and Ingrid took with them when they moved to Raleigh, where Denis would take up a university career (he had been teaching in a high school). There they bought the Mies van der Rohe coffee table (65) and chair (53) to complement it (they also bought the first Palaset cubes [26]), an ensemble they moved all but intact to the room it now occupies. In each site the couch had claimed pride of place, either focal (as in Worcester and various arrangements in Cleveland) or central (as in other arrangements in Cleveland, as in Raleigh). The couch thus combines in itself the authority of furniture, the prestige of heavy investment, the perfection of the white, the power of the center.

Nonetheless, its situation is not simple. In fact, it is embroidered with contradictions. This symbol of propriety is a (potential) site of indecencies: at the very least it is an open invitation. This embodiment of the useful undermines its utility by dressing in white: the sensibleness of middle-class realism would dictate a color that . . . *hid dirt*. This blazon of the bourgeoisie takes the form of the modernist polemic against the bourgeoise: there is no tufting, the feet are not turned, the arm is not scrolled, there is no crest rail, no rosette: there is only the . . . *necessary* (the clean low lines, the trim foam cushions for which Paul McCobb was known). These contradictions involve themselves in further contradictions. The bourgeois fat trimmed away by Paul McCobb returns in the "conspicuous austerity" of the white slipcover. Yet this is also less than it seems, for the slipcover actually makes *fewer* demands on the economy of the clean than scrollwork (which gathers dust), fine woods (which have to be polished), and upholstery (dry shampoos); and in any case middle-class realism (the couch will be dirty) is in perpetual conflict with middle-class idealism (the couch should be clean) in the first place. In a similar fashion, the indecency of the invitation extended by a bed in the living room is canceled by the practical advantages of having an extra bedroom (that just happens to be at the foot of the stairs). In this way the room is given as many roles in the home as the couch plays in the room: it is a bedroom (someone sleeps over on the couch), it is a sickroom (someone sick spends the day on the couch), it is a lounge (someone is sprawled on the couch), it is a living room (two are companionably talking on the couch), it is an office (a student on the couch, Denis in the wicker rocker), it is a study (the couch light is on, Denis is grading papers, the kids are reading a book [Ingrid is in the kitchen cooking]). Driven by the amphibology of the couch, the living room unfolds the extent to which the functionalist divisions among room presumed to mark our distance from the past are less than real, the extent to which the Weberian distinction between *place of work* and *place of residence* remains an ideal. The living room is *not* a seventeenth-century Parisian *chambre*, dominated by a bed, but in which demountable tables could be erected for dining with company of whatever quality; but neither are its functions as narrowly drawn as we pretend. Our room is entrance hall (there is no vestibule), reception hall (site of introductions, small talk, where decisions are reached to stay, or pass through the room [which is the only hall] to the kitchen or the back yard), parlor (for instance, when Denis's or Ingrid's parents visit), occasional dining room (especially at Christmas, but lunch is not infrequently eaten here by adults and their guests), music room (this is where the best speakers are, the records), dance hall (in which case, uniquely, the couch is pushed against the wall), playroom (especially at Christmas, but not uniquely), to say nothing of, as just noted, office (nor is it merely students with whom Denis works here), study, lounge, sick room, bedroom, and so forth and so on, in most of which the presence of the couch as divan, chair, bed, room divider, desk is taken for granted (it is the moving of the couch that makes dancing in the living room slightly . . . *daring*). What is essential to grasp here is not that these roles . . . are multiple (inevitably a form of wealth) but that they are contradictory (invariably a loss of control): from within any given role another is often impossible to entertain: *we don't do that in the living room.*

Ah, but we do.

The question raised by each role played by the couch is: what kind of people are these using it in this way? The Woods want to be the kind of people who have bedrooms *apart* from the rest of the house ("no one can be close to others, without also being able to be alone"), but they also want to be the kind of people who can put their friends up overnight (they are the lords of the castle). Inevitably these will conflict and someone will have to sleep on the couch *(a bedroom will materialize in the very core of the home)*. Or they want to be the kind of people with a place to retreat to that is always neat and tidy (the living room), but they also want to be the kind of people who can just kick off their shoes (they are not stuffy burghers, rigid aristocrats [Denis's mother's father once dismissed a maid for coming shoeless to the table, his father's father was shocked when his mother answered the door in bare feet]). Inevitably these will conflict *and all too often they will enter the room to find the newspaper scattered on the coffee table, shoes on the floor, and the couch cushions in disarray* (someone has been playing on the couch [maybe even Denis wrestling with the kids]). Or, yes, they want to have a place to receive guests, but they also want to have a sick kid downstairs if he wants to be. Inevitably these desires will conflict, *and amid a cloud of explanation* (all perfectly well understood) *the guest will be conducted to the kitchen.* There are not enough rooms (there are too many needs): it is all so commonplace as to scarcely merit description, except that neither the kind of people they are nor their ability to realize them are independent of their social and economic capital. *We're not the kind of people who entertain in the bedroom* depends on a certain reading of the bed and a commitment to sexual discipline as well as the possession of other rooms in which to entertain. As means shrink or expand, so do distinctions and the ability to manifest them in action (class fractions are completely defined by the appearance or disappearance of distinctions and the order and rate in which they do so). Here's an example: Denis and Ingrid have just bought a futon and frame for the "middle room" upstairs (which is not to call it a bedroom [it is Ingrid's "sewing room" {but Denis sleeps in it half the time (Ingrid snores)}]). This can be a couch by day, a double bed by night: *expanding means*. Not surprisingly it expanded the . . . *couch function*. As the piece of furniture most supportive of all those functions gathered together under the name *living* (couch = living room), the couch is the piece of furniture most completely implicated in each of these scenarios.

The rules catch this importance (more rules encrust the couch than anything else in the room) but also this confusion of purpose, of meaning. Certainly there is great concern with the appearance of the couch (which *is* white [and whose slipcover does have 250 snaps]). *Don't mark back with shoes, no shoes on the couch, don't put food on it, don't throw up on it* are no more than particularizations of Denis's *don't get the couch dirty*, which in the uttering marks the incompatible uses of the couch (the show piece that insists on being white is not the sickbed on the ground floor for convenience and company). The heavy investment protected by *don't throw yourself onto the couch—sit down on it, don't sit on the arms, don't sit on the back,* and *don't hang off back of couch* is not the jungle gym implied by *don't climb over the couch, don't fight on the couch, don't hide things under or behind the pillows,* and *don't take up the pillows and sit underneath them* (a rule the kids violated to illustrate). To an unrivaled degree, the couch embodies in an inescapable fashion the contradictions of the society that produced (and reproduces) it. The couch is a sign whose signifier perpetually eludes the monogamous union sought by its signified. No alibi serves to save the couch from its lies (its utility is completely denied by its excesses). Because it nonetheless (or: therefore) *is* the piece of furniture required for a living room *to be* a living room, the battle over taste waged by the rising and declining, managerial and academic fractions of the middle-class will inevitably take the couch as their battleground, as conflicting generations will signal their solidarity by agreeing on the vehicle through which to articulate their differences. Whether on tufted cushions and turned legs or unobtrusive legs and foam cushions . . . *the couch will be sat on.*

And why not? None of this is to deny its *comfort* (there is no lack of reinforcing animal pleasures).

It is only to insist that there is more to comfort than ergonomics.

Peter's Box

From a distance it looks like a small window punched through the wall to another world, maybe the Middle Ages. But up close you realize it's a shadow box, glazed against the dust; and far from the Middle Ages, it's a modern world you wished you didn't know.

The box itself, not quite eleven by fourteen inches—two deep—is carefully con-

structed, but from a wormy wood of pronounced grain. Painted a flat white, it could be a cousin to one of Joseph Cornell's. Floating within, maybe half an inch from the back, is a crude piece of some artificial construction material not seven inches by ten, brutally hacked from a larger sheet. To it has been affixed a peculiar drawing of what looks like one of Charles Babbage's calculating machines, done in ink on a piece of cheap lined writing paper stained or painted and faded to the blush of a summer's eve. Below the machine is a postcard sun-bleached to shades of cyan: in the foreground a system of terraces is linked together by stairs; behind this an old church; and behind this and off to the side, modern high-rise apartment slabs. Maybe you recognize it as the Plaza of the Three Cultures in Mexico City, or maybe the name, Tlatelolco, in crude blue letters at the top of the sheet, gives it away. The postcard is framed with thickly applied red paint on three sides, a dark green along the top where wires from the machine meet the urban scene. The paint has been dusted—or daubed—with what looks like powdered copper. Just above the machine a narrow strap of leather—ripped at both ends—has been glued. Dangling from one end by an inch or so of twine is a tarnished silver arm. This is a *milagro,* one of the votive offerings found for sale in the stalls outside large churches in Mexico, but even if you don't recognize this, it is evidently not something from a charm bracelet.

Or maybe it *is* the Middle Ages: they at any rate made no claim to reason.

rule 145. "Don't touch—just look at it" (Chandler, kCON, tPRO).

What is this?

Whatever it is, it is *not* a gem. Chandler makes this clear. Gems gleam, gems sparkle. *Keep your fingers off the windows* (RULE 26), *don't smudge up the windows* (RULE 27), *don't breathe on it* (RULE 28), *don't let your friends put their noses on the window* (RULE 32), *don't wipe off with your hands. You only smear them up* (RULE 33), and *same as all pictures—don't put your hands on it (same as windows)* (RULE 90) are each in the service . . . of the shining. When Chandler interdicts putting your hands on the Ernst *(same as windows),* it is the Lucite whose appearance he is guarding, not the lithograph he fails to even notice (20). But when here he interdicts touching, it is with which part of the body *should be* involved that he is concerned (it should be the eyes), not the glass which here he no more than . . . *looks through.* That is, in the space between *don't breathe on it* (RULE 28) and *don't touch—just look at it* is inscribed the difference between glass as signifier and glass as irrelevancy (though even here it unavoidably marks Peter's assemblage as precious), that is, between glass to be looked *at* and glass to be looked *through.*

Peter would approve. Because no matter how pretty his construction is—and there is something about it of a miniature by the Limbourgs—it is meant to be . . . *looked through:* it is what it is about that *really* matters, and what it is about is death.

Tlatelolco is a name that deserves to be printed in blood, this despite the fact that originally it was no more than the name of an island in Lake Texcoco. A hundred and fifty years after its occupation (by disaffected Aztecs), it was subjugated by and absorbed into its neighbor in the Valley of Mexico, Tenochitlán. In this great city its plaza was to become the site of that market whose abundance and variety would so bedazzle Bernal Díaz del Castillo half a century later. It was also to be the site where Cuauhtemoc would make his final stand—the blood of whose defeat saturates the soil from which rises today the Plaza of the Three Cultures (Aztec, Spanish, Mexican)— and where, in 1968, hundreds of students would be murdered, gunned down like those in Tiananmen Square, by soldiers who then would rampage through nearby apartments raping and killing all they accused of sheltering students: how many died? Who knows. But the official number was lower by degrees of magnitude than those of eyewitnesses, one of whom was Denis's brother, Peter. "I was there," he has said, "and I wanted to express it in my art." Though several paintings would follow, the box was his first attempt: "I found this drawing out in back of a building of a machine. I decided it was some sort of modern murder device, an electric chair or something. And here is a picture of the Plaza of the Three Cultures—where there were earlier massacres, where there were massacres in ancient times. . . . "

At the very least, then, the box is some sort of *memento mori,* not merely a souvenir (a remembrance, a memorial). Modern murder devices are everywhere (Denis and Ingrid live only yards from the North Carolina State Central Prison with its literal electric chair) nor are they necessarily *intended* to kill (though there are more than enough with this exclusive aim), and Denis cannot think of this apart from the knowledge that in the United States alone *cars* kill 50,000 people a year. But Peter has added further elements, a piece of leather from the thong of a staff from Chamula in the highlands of Chiapas in the south of Mexico "for its magical associations" and the *milagro* "so that the whole thing is a *retablo.*"

That is, a *reredos,* an altarpiece, one of those pictures found throughout Mexico above and behind an altar, originally intended to *instruct* (by illustrating scenes from the lives of the Holy Family, apostles, saints) but after the Counter-Reformation to actually *arouse* religious feeling (with increasingly expressive portraits, with images of uniquely vivid events). So, far from being merely an object of art—though it is expressly that—it is almost a cult object tied directly to the Indian (through the piece of leather) whose initial oppression in Tlatelolco is the archetype of each subsequent act of oppression . . . *including the one to come.*

Does Chandler understand this? Probably not in this way, though he has been present often enough when Denis has explained it to others. But even if he did, to acknowledge this would be no more than to nod toward the significance of the box as an isolated thing. But it is not isolated. It is caught up in interconnected circuits of meaning that if followed to their ends would embrace everything in the room. We need not go that far, but if we are to understand the values implicit in the rule, the most important routes into the rest of the room need at least to be indicated. We can see seven of these, the first of which is doubtless that of family: the box was not made by an unknown Artist (visions of an old man with white hair like the God of Denis's childhood friends), but by *Chandler's uncle,* one moreover who has painted the portraits so prominent in the room of Malatesta (41) and Genovevo de la O (64). The second is that of Mexico, pulling together not only the box and the Ferris wheel (25), Lacandon drum (39), king (55), wooden car (60), portrait of Genovevo de la O (64), and baseball game (68), but the lives of Denis's and Ingrid's parents (the Woods lived in Mexico on two occasions and traveled there often, Ingrid's father sold chemicals there and the Hansens had extensive acquaintance in the country) as well as those of Denis (frequent visitor) and Ingrid (who carried out fieldwork with Denis in the highlands of Chiapas). The third is that thematics of proletarian revolt that runs through the Ernst (20), the box, and the two paintings (furthermore, these are united with the folk art [including the Cherokee basket {24}], certain records [27], the uncurtained windows [17, 47, 50] and unlocked door [7] in a broader populist theme). The fourth connects the box with the Ernst—through his collages, the collages (32, 33), and the wire on the mantle (63) in a thematics of the *trouvaille* (linked in turn with the wicker rocker [35] and many of the records [27] in the broader themes of "the found" and "the used"). The fifth picks up the "magical associations" of the piece of leather and the *milagro* and observes their alliance with the Ernst, the Lacandon drum (39), and the king (55) in a thematics of "the savage heart." The sixth observes the explicit relationship of the box *as collage* with the collages (32, 33). The seventh notes the definition of this collage of found and magical objects as a *work of art*—which is what Peter set out to make and how Denis and Ingrid treat it—and thereby makes the connection to *modern art,* that is, to the Ernst (20), Denis Woods (32, 33), Rozzelle (37), Yunkers (40), Strand (56), and other Peter Woods (41, 64).

It is its suspension in this multidimensional web of associations, endlessly ramified, continuously—and off-handedly, casually—referred to, that Chandler internalizes and out of which he will construct the (doubtless very different) meaning of the box in his life. Denis has told him about the slaughter, and the box's images and gestures are not inaccessible (the severed arm, the slathered paint), but Chandler has never been to Mexico, the box was made by his uncle, not his brother, and modern art will not be for him what it was for his father. Nonetheless, his rule offers the box a respect it withheld from the Ernst.

Death has entered the room: everything . . . is more serious.

Denis's Collage

To see the room wheel across the mirror of its glazing is to appreciate in the space of the collage your movement across the room: the windows behind the wicker rocker balloon within its frame, in a glance the doorway lace implodes, from the corner of your eye, the final flare of the little window by the stair.

To look at the collage is to stop this dance, to see past the phantom room to the ghosts within, for the collage too has been bleached by the sun and it is all a question of chalks and alabasters, lilies and milk. In the foreground a man in a shooting crouch aims a camera at a spaniel. Except for the camera and a pair of Jockey shorts the man is naked, though more at home than the dog on the rip-rap of a dock on the coast of Maine. From above, a giant orchid swoops into the frame. In the background small sailing boats are swallowed in a fog equal parts Maine mist and Carolina sunlight. Though the frame is seventeen inches by twenty-one, the Maine scene is a little less than ten and a half by thirteen and a half, the trim size of the old *Holiday* which is where it came from. The guy was snipped from *Sight and Sound,* orchid and dog from *Life.*

They are part of it too, mat and frame, glass and glazier's points, screw eyes and picture wire, picture hook and nail. The pretense is the images float on the wall, but the truth is the usual story of a crabbed technology unfolded over time. Caught up here in invisible support of some *papiers collés* are minute but not negligible quantities of alumina and soda, silica and pine, magnesia and lime, hematite and coking coal, the latter from those fields in West Virginia where still the spirit of John L. Lewis can find no rest.

hand-off   2. "Same as every painting" (Chandler, Randall, to RULES 90, 91, 145).

We have not characterized the glue out of deference to Max Ernst's dictum *ce n'est pas le colle qui fait le collage,* but in fact if you look closely along the edges of orchid, dog, and man, you can see the fine bead of rubber cement that is the invariable consequence of its use. Its presence here indicates a collage made at least twenty years ago, when Denis forswore its use, and in fact this one, *"What Are the Wild Waves Saying, Baby?" II,* was glued up in 1968, just a few years after he started making

32

collages. Twenty years later, in the catalogue to a show in Raleigh's Municipal Building, he had this to say about them:

> When people ask me what the point of my collages is, I usually turn it around and ask them what they think it is, because certainly that is a big part of the point of *showing* the collages, to provoke thought.
>
> But it has little enough to do with why I started *making* collages a quarter of a century ago, or why I continue to make them today. This fundamental reason is quite simple: I make collages to free myself from the tyranny of the images which constantly assault me from magazines and movies, newspapers and billboards, a freedom I achieve by invading the images themselves, cutting them up, and reassembling them.
>
> All of the images that comprise the collages in this show were once imprisoned in the contexts in which I found them, were slaves in the service of messages in which I had scant—if any—interest. The olives, for instance, had been snagged by an advertisement for snack foods; the belts were trapped in a mail order catalogue; and the young baseball players were being held hostage in a public relations campaign for an oil company. Unless you looked carefully you were unable to see that the olives were just olives, that they had nothing to do with "snack foods"; that the belts had no interest in being bought or sold; that the kids themselves had nothing to do with this oil company, that their smiles and excitement had to do with being young and healthy and maybe winning ball games, or at the very least with having fun in a photo session for which they were being (well) paid. Once you see this, though, the advertisement or the scene or the editorial illustration ceases seeming innocent. You recognize it for the jailhouse it is, and suddenly you feel obliged to hack your way in with a pair of scissors to set the kids free. To liberate the belts. To save the olives.
>
> Relocating them in another environment—itself released from the prison of its identity, its context—enables you (me) to repoeticize the world, to rescue it from the reductionism of an exclusively utilitarian rationality.
>
> And if you imagine I have merely tossed these images into another prison, one of my own devising . . . I won't argue with you. Instead I'll encourage *you* to pick up the scissors, encourage *you* to set the images free again. For it is not the least of its attractions that this art is proletarian, that anyone can do it, that— liberating—it is also democratic. In this way, at least, it practices what it preaches, and that also is the point of these collages.
>
> Cut, any of you, and set yourself free!

The departures all but . . . *announce themselves:* as *work of art* to Ernst (20), Peter

Woods (31, 41, 64), Denis Wood (33), Rozzelle (37), Yunkers (40), and Strand (56); as *collage* to the other collage (33), to Peter's box; via the *trouvaille* to the box (31), the Ernst (20), and the wire on the mantle (63), and with a change of trains to everything caught up in "the found" and "the used" (for example, the wicker rocker [35]); via *the dream* to the Ernst (20), and the other collage (33), the Rozzelle (37); via *proletarian revolt* to the Ernst (20) and the box and paintings of Peter (31, 41, 64), and, with another change of trains, to everything caught up in the larger populist theme (for example, the unlocked door [7]); and finally (but not conclusively) via *family* to the living that is the home (in which the room is an organ). Here we observe the production of the collages—of which ours is no more than an early example—in which the family is necessarily swept up: in: the cutting out ("wouldn't this make a good collage?"), the assembling (Denis returns to his work table to discover everything rearranged, by Ingrid, by the kids), the reviewing (the collages are laid out on the dining room table: everyone comments), the gluing (Denis is invariably hysterical), the framing (the whole downstairs is taken over, glass washed in the kitchen, frames assembled in the dining room, framed collages in the living room, everywhere), the opening (the polite conversations, the friends, the acquaintances, the frosted grapes), the reading about them in the newspaper (again: everyone comments), the putting away (they are stored in a closet in the kids' room), and getting out (to give away, to sell, to look at).

Evidently it is not the glue that makes the collage (the room is not glued): it is the labor, it is . . . *the living*.

For the kids, this labor is an ineluctable presence *in the collage,* a quality of *having been made* that is thereby encouraged to penetrate the room through its alliances: if the collage is a made thing, so are Peter's paintings (41, 64), Ron's drawing (37), Adja's lithograph (40), so is . . . *the room* (which is a collage).

Denis's Collage at the Foot of the Stairs

Ruddied mirror, sudden green at summer's edge, black and white at night where the street lights throw the shadows of the window on the glass . . .

Yet from the stairs it is a door cracked on another universe, where a dark red carpet lies on a desert floor and a monstrous fork and scarlet tomatoes dwarf the runner that approaches in his yellow shorts. At world's edge a chink of sunlight, flaring into dawn.

Nine inches by twelve, this collage is lodged in a metal section frame sixteen inches by twenty. The anodized aluminum plays with the light like the glass.

hand-off 2. "Same as every painting" (Chandler, Randall, to RULES 90, 91, 145).

Because we have concluded a discussion does not mean there is no more to be said about its object. Each of these disquisitions is no more than a sign with an arrow pointing in a certain direction: each connection observed is no more than that, there is little notion of its modality, less of its strength. With respect to the collages one could have asked, for instance, what it is that in freeing himself from the orthodox stupefactions of movies and magazines, billboards and newspapers, Denis hopes to accomplish. In a talk he gave in 1985 to introduce a retrospective of his collages, he put it this way:

> I want to say that they're not merely *like* my life, they're not merely a lifelike product that I happened to throw up out of my life, but that they're part of it and generative within my life. They free me. I was always—I still am—caught up in the trap of social identities, who I was, who I am now, human being, male, husband, father, college professor, Ph.D., intellectual, with-it, dignified. I am burdened, in fact, with a litany my mother made for me when I was an elementary school student running for Fifth Grade class president or some similarly titanic position. And she made me a list of words the first letters of which spell my name. And it goes: *D*emocratic, *E*nergetic, *N*eat, *I*ntelligent, *S*ensible, and that spells "DENIS" . . . Wait, we haven't got the "WOOD" yet: *W*orthwhile, *O*rderly, *O*riginal and *D*ignified. Dignified. That's the one that has always bothered me.
>
> And all of these attributes—and all the rest of the attributes that are much more serious than these attributes—all these attributes, all these identities, trap me, they force me to consider, "Is this the way a Ph.D. behaves?" "Is this the way a college professor behaves?" "Is this the way a human being behaves?" "Is this a decent human thing to do?" Max Ernst, writing about his collages, observed that:
>
>> A ready-made reality, whose naive destination has the air of having been fixed, once and for all (a canoe), finding itself in the presence of another and hardly less absurd reality (a vacuum cleaner), in a place where both of them must feel displaced (a forest), will by this very fact, escape to its naive destination and to its true identity; it will pass from its false absolute, through a series of relative values, into a new absolute value, true and poetic: canoe and vacuum cleaner will make love.
>
> And, of course, that's how it is with the collages.
>
> Reflecting on Ernst's collages, André Breton added: "Who knows if, thus, we are not preparing to escape someday the principle of identity?" And Max Ernst and André Breton felt that the work that they were engaged in, which they thought was work, and important work, labor—of the kind I do when I make my

collages—was enabling them to escape from the trap, was enabling others to escape from the trap, enabling them to no longer worry about, "I can't do that, that's not the way a ——— identity operates, or behaves, or dresses."

And obviously this is the last thing I want to say about my collages, the way putting them together forces their parts to shed their everyday identities, and makes it clear that I, and you presumably, can do the same.

But whatever "you" may have managed along these lines, it is evident that Denis has achieved little enough: the 223 rules volunteered by the family Wood for no more than the living room testify *minutely* to the limited realization of these surrealist desires. Moreover, they testify how far Denis will go in order to *reproduce* his identity; for what—after all—*are* the rules but a machine for making Randall and Chandler *d*emocratic, *e*nergetic, *n*eat, *i*ntelligent, *s*ensible, *w*orthwhile, *o*rderly, *o*riginal, and *d*ignified? For making them, that is, d-e-n-i-s-w-o-o-d? For making them, *Denis Wood?*

What else are the rules? They are a machine for making Randall and Chandler . . . *Ingrid Hansen!* Because they are burdened with this double imperative, they are incapacitated from achieving either, producing instead an, as it were, unanticipated anagram (of some amalgam) of their names, as though the code of a kind of cultural genetics could be read there. None of this may be allowed to surprise us, not even that Denis's mother thirty years ago was permitted to anticipate in her acrostic so many of the terms in that system of values the rules have enabled us to unveil. Is there any we have not already encountered? *Democratic* certainly (theme of populism and so on); *neat*—and *orderly*—above all others; *sensible* (is this not the Voice of Convenience?); *worthwhile* (unattainable ambitions, high culture, paintings); *dignified* (the horror of animal abandon, the Voice of a Certain Easy Formality). The syntagm that results (which we see *is* Denis, turned through a certain angle) may not be taken for granted: it is a combination of character traits valued by his mother and encouraged in him at the expense of others. Another mother would have made *her* Denis a different slogan (*d*ynamic-*e*xciting-*n*ecessary-*i*ntense-*s*tellar, perhaps) or none at all or not allowed him to participate or kept him home from school . . .

We see well enough even as the rules disclose the room's failure to live the implications of its collages (though to avoid exaggerating this, it is salutary to reflect on the rooms of Ernst and Breton), that the collages expose the extent to which the room is first and foremost a field of rules whose end is all but invariably its own reproduction, as the end of Denis is his reproduction, and the end of Ingrid is her reproduction, in Randall, in Chandler, in the rooms they will live. Because this field is incompletely determined—it is like genes, subject to a double imperative—the whole system will illustrate the Marxist dictum that, though men will make their own history, they will not make it as they please, but under conditions inherited from the past.

This is a no less perfect description of collages, made not ab nihilo, but with stuff cast on the beach of the present by the ocean of the past (this is variously deep).

The room? Collage.

XXIII · THE FIELD (AGAIN)

We know that kids in a room are like particles in a force field: everything of the room—door, Cherokee basket, couch, collage—is capable of contributing to the field.

But not everything does.

Whether something is "turned on" or "off" is, as we know, a function of the degree to which the values latent in it are threatened (by the kids, by the delivery man, by the neighbor's dog [by the barbarians]). When these values are threatened, they are transformed into rules (by Denis and Ingrid, by the parents [by those responsible for living the room]). This implies that there are two fields in the room, (1) an implicit *field of values and meanings* participating in those of the world (because the room is an open system); and (2) an explicit *field of rules* (which we are using to gain access to the former).

These two fields correspond to the two types of behavior manifested in the room. A mother and small child enter the room. The mother bends to admire something—say the wooden car (60)—touches it, may even pick it up. Later the child approaches the car, squats before it. Suddenly: *oh no, dear, that's not to touch—here, let Mommy find you something to play with.* This is invariably followed by the *sotte voce* plea: *is there anything she could play with?* meaning, of course, anything whose latent value would remain unaltered under barbarian assault (an old pan).

In the room, the mother moves through the *field of values and meanings.* For Denis or Ingrid to have *expressed* the rules to the mother (RULE 204, *we shouldn't wind it up and play with it*) would have implied that she didn't know better (would have implied that she saw in the car a toy instead of folk art) or that she were a child (and didn't know better [that is, saw in the car a toy—which it is—instead of folk art {a category of analysis not present in the barbarian repertoire}]). In both cases Denis and Ingrid would have patronized the mother, violating their own sense of values and potentially alienating their guest (who is a jewel on the cushion of hospitality). For Randall or Chandler to have expressed the rules would merely have been *cute:* the play civility of little barbarians. (Here the sign of the barbarian is the failure to heed the rule forbidding the recitation of rules to the already civilized.)

The child, on the other hand, moves through the *field of rules.* Just as the values and meanings did not have to be unfolded for her mother, so the rules do not have to be explained to her. From the moment she roused into consciousness that morning it was a consciousness of what she must and must not do, and here, in this room—where she has never been before—what appears to her mother as an attractive object (as, that is, the center of a field which attracts her, which pulls her in, which she stoops to touch) appears to the child as the center of a field which says: *don't touch.* Why should this be? Because kids (in general) are discouraged from touching things they are not explicitly permitted to *(you can play with that, don't touch anything else)*. Parents are thereby (more or less fearlessly) enabled to take their kids: out (to department stores, to the homes of strangers). But also because (more specifically) living rooms are those rooms most likely to be pervaded by this rule, its variants and (obsessive) particularizations (unless preempted by the still more heavy-handed *stay out*). Within this room (more specifically still), it is above all others objects in postures of wanton display which have been interdicted, and it is, as we have seen, precisely their lack of utility which heightens their value (it is through the resolution of this paradox that the child will become a connoisseur). Though she has never been in *this* room before, she has, therefore, lived its structural analogues: precisely as the child learns to speak sentences she has never heard, so she learns to live situations she has never experienced. Even this one, intended to embrace the child as a guest no less than her mother:

"I thought I told you *not to touch.*"

"Ah, Mom, I'm *not* going to hurt it!"

"It's okay, she can play with it."

"Well"—hesitantly—"if you say so," where the hesitancy is for the child the certain sign of the limitation of this extraordinary privilege (it does not go beyond the door).

Evidently, then, the rules do not need to be uttered for a field to be present: they need . . . *to have been uttered.* Rules are not the natural associates of values and meanings: these must be transformed into rules by those responsible for them. *This transformation will take place whenever the values and meanings latent in an object are threatened.* Any threat will "turn" an object "on." This will initially be accompanied by the emission of a rule *(don't you EVER touch that).* It is this emission of the rule that alerts the child to the perception of his behavior as a threat by the rule-emitting adult and which causes the object to "glow" with the force of the rule. Any other object identified by the child as similar will similarly "glow" with the force of the rule (with respect to which identification some kids will be strict constructionists *[you said THAT chair]*, others more liberal *[I thought you meant EVERY door]*). In this way the room becomes a space "shaped" for the child by the fields "turned on" through the emission of rules prompted by his or her real or apparent threat to the values and meanings implicit in the objects that generate the fields. This knickknack shelf populated by china shepherdesses and fragile jars filled with a potpourri of roses REPULSES the child with a force proportionate to that conveyed in the emission of the rule. Forced to the side, the child approaches the claw-foot mirror-topped coffee table with its gilt candelabrum. This PUSHES him off at an angle toward a side table with a pendulum clock in its crystal case which SHOVES him into a carom that . . . slips the kid through the door. What he follows in this way is the ordinary path through the room, but his path is not "natural" (it does not result from trying to move on a sloping floor through a gravitational field). It arises every time it is attempted, as the resultant of a system of intentions (what the kid is going to do [he is like a charged particle]) moving through a field of rules. Each object along the path glows with a repulsive force that shapes for the child the space the room is for him. It is a lot like a pinball layout, but it is really a field laid down by the rules. The field of rules creates . . . a hodological space.

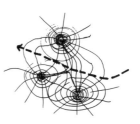

Of course it is not so simple. The room is also a space shaped for the child by the field of values and meanings which are largely (though not exclusively) attractive. This follows from the fact that what is for the child a field of rules is for the adult a nest of comforts no less desirous to the child. The potpourri of roses PULLS the child to the knickknack shelf, the mirror on the table SEDUCES him to bring his nose as close as he can, the clock locks the kid in a TRACTOR BEAM ("Why are we still moving toward it?" "We're caught in a tractor beam and it's pulling us in." "There's gotta be something you can do about it." "There's nothing I can do about it, kid. I'm full power."). Such attractions are not simply "erotic." Many—in other rooms, most—are powerfully instrumental, and this instrumentality is often straightforwardly "utilitarian": door, window, floor, stair, banister, chairs, couch, table, lamps *afford* support (as the table supports a cup), light (streaming through the windows, puddling around the lamps), access (through the door, up stairs). These *affordances* "permit" the living of the room: the afternoon papers arrive for Randall's route, he brings one in, plops it on the table, sits on the couch, reads (now everything, but originally mostly the comics). Denis, Ingrid may join him. This is a *routine* composed of numerous behaviors that has evolved over time. It is a routine evidently attracted to the living room (the living room draws this routine in [as it has never the unpacking of the grocery bags, analogously attracted to the kitchen]), and this at the most fundamental level, for the reasons that the living room *affords* proximity (the papers are delivered to the front door), seating (two of us on the couch, a third on the wicker rocker), light (the windows in the summer, the lamp by the couch in the winter). These *affordances* we wrap up in the notion "meaning": providing seating for reading the newspaper is one of the things the couch "means." At the same time the routine harmonizes with others we associate with the living room: it involves reading, thinking, discussion (explanation of the political cartoons [originally confused with the comics]), it is an art *(contemplation of the news),* it is a routine of consumption (oriented toward the front of the house), not production (even writing is produced toward the back— Denis's study—or upstairs). These significations we wrap up in the notion "values": providing a place for the contemplative arts is how the living room lives its values. In this way the meanings and values incarnated in them make each object in the room attractive, gives each the power to draw the child to it. Randall with the paper is *sucked* down to the couch, the paper is *pulled* to the table. What results is an ordinary routine, but this routine is not "natural" (it does not result from the action of sunlight on cellulose). It arises every time it is attempted, as the resultant of a system of intentions (Randall is a charged particle: he intends to read the paper) moving through a field of values and meanings. Each object involved pulsates with an attractive force that shapes for the child the space the room is for him. The field of values and meanings results in . . . a hodological space.

These two spaces—that arising from the field of rules, that arising from the field of values and meanings—are not identical, nor do they differ only by being the "nega- tive" and "positive" sides of a common form (that of the living of the room). They are frequently in conflict and suspend the child in conflict. The table by the couch affords support equally to paper and feet, but the former is permitted, the latter interdicted. In this way we can see that not every possibility latent in the field of values and meanings can be realized by a child simultaneously in motion through a field of rules. Though, taken together, the two have a powerfully *pre*scriptive effect, the rules for the Woods— and we believe most others—are deeply *pro*scriptive: *you know the rules: you're not supposed to fight on the couch.* Rules do not model behavior: they set its limits. This results in an essential efficiency: *it is easier to say what a couch is not than what it is* (it is easier to say not to fight than how to get along [what does it mean to love your brother? {but what it means to kill him is easy}]). This in turn contributes to a powerful freedom: *everything not forbidden is implicitly permitted.* This is not to deny the existence of endless routines and habits, standing behaviors and "jobs" (the Woods' word for "chores") that can be characterized in positive language, but it is to insist that such descriptions are unendurably "artificial" and usually *after the fact.* Randall may ask Denis or Ingrid to tell him how to string the lights on the Christmas tree (even this is better learned by showing than telling) but no positive "rule" structure even exists to describe reading the paper, visiting with Ingrid's brother, talking with friends, playing a game with Den or . . . unwrapping presents: while this evergreen event evolves, *the rules hold*. Kids may or may not be guests, but still *there is no fighting on the couch;* the presents may be piled above or beneath the table, but there will be *no feet on it* just the same. Like every routine, unwrapping presents takes form over time through observation, mimicry, role-playing, discovery, and invention, but unvaryingly within a wrapping of rules that interdicts transgressions of previously threatened values and meanings. *No rules will exist where these have not been threatened.* From this follows the paucity or complete absence of rules for Peter's box (31), Denis's collages (32, 33), Ron's drawing (37), the Yunkers lithograph (40), Peter's paintings (41, 64), the side screen (51), the side window (52), the Strand photograph (56), the fireplace plant (58), the plaited fan (59), the turtle (66), and the plant on the table (67). The plants may be subject to a hand-off (and it would not be unreasonable to assign one to the art work), but no rules were collected from the Woods for these things because up on the wall (the art work) or out of the way (the window, the fan) the values latent in them were never subject to threat. With his collaborationist's instinct Randall had insisted that these absent rules were merely unwritten, unspoken, but Ingrid observed that they were never specified. And why should they have been? It is in keeping with another efficiency that rules emerge . . . only when needed.

If the life of the child could unfold in perfect consonance with the values and meanings of his parents, there would arise no reason for such rules: the space created by the field of values and meanings alone would provide unfailing support for action which would be all but automatic (the child would be a clone). Profoundly authoritarian households do exist in which every action is positively modeled and where what is not permitted is expressly forbidden. But the concomitant inefficiencies and restrictiveness necessitate a value system correspondingly spare and simple; and the consequence for the child is an experience of devastating oppressiveness despite the absence of *don'ts*. In such a space there is no room to breathe.

To touch or not to touch . . . The freedom of the child is suspended here, *between* the spaces. This is no reference to the Christian praxis of an innocence certified only by temptation, but to the reality of the unendingly nuanced negotiation that stands behind our caricatures of attraction and repulsion. It is in his original response to the fields, breaking this rule—but following that, ignoring this attraction—but succumbing to another, that the child forges his being. It is never a question of his being forced (though he may describe it in exactly this way) but of solving the equations relating his intentionality (his charge) to that of his parents (the fields they implace). Never contemplative, each attempted solution achieves a closer approximation to an answer which is endlessly reintroduced as a new condition (this is what it means to live

34

the room). With such increasing self-consciousness . . . he comes into his own. Does this mean the fields are not real? No: it means it takes two to tango.

The Couch Lamp

Eventide yellow is the lamp's light, matin's cool the chrome of neck and base, an evanescent elegance sheathed in a honeyed linen . . .

The honey comes from age. The paper and linen of the shade have been subjected for nineteen years to the degrading effects of their own acids, of light and heat, of cigarette smoke and sulfur dioxide, fading, discoloring, turning brittle. The shade holds its shape thanks to the rings at bottom and top (there are no ribs), the one at the top tying it to the finial through the spokes called the spider. The finial crowns the harp, which slips into the harp sleeves, which are forced onto the harp wing, which is fastened to the neck with a set screw.

The light comes from electrons. These are falling back to their proper orbits around the tungsten nucleus from which they have been cattle-prodded by the electric current zapping through the filament (which will reach 4,500°F [don't touch]). The filament is suspended in an argon gas (this is to prevent its combustion) caught in a lime glass bulb. It is held in place by support wires, and connected to the tip and ring contacts of the base by lead-in wires (these are all encased in the glass stem rising from the base). The base is screwed into the socket, whose shell—penetrated by an on-off switch— slips into the socket cap that is screwed to the nipple that passes through the harp wing to the neck. Within the socket, ring and tip press against contacts soldered to the leads of stranded copper wire that, sheathed in an insulating plastic, passes through the neck of the lamp—a chrome-plated steel tube screwed into the base—and base—a heavy casting cloaked in a chrome-plated steel sphere truncated at the floor—to slip out through a plastic fitting. The wire snakes across the floor to the baseboard socket (behind the little record cabinets [23]) where it makes contact with a plug. This is another pair of contacts and a pair of protruding blades or pins encased in plastic.

The electric current exciting the tungsten electrons to perform their light-emitting acrobatics reaches the baseboard socket through the house wiring, fed, through head, meter, and box, by the 240-volt drop from the secondary in the alley. This droops from a transformer three houses up which is fed in turn by a 23,000-volt line from the Method Substation a mile or so to the west. This is fed by a 230,000-volt line tied into the grid, though most of the electricity powering Boylan Heights slips along it from the coal-fired generators run by Carolina Power and Light in Roxboro and Goldsboro. The coal reaches them in the black-and-white Norfolk and Southern hoppers that slide down from the West Virginia hills under little more than the force of gravity. Which traces the eventide yellow back to the sun, the deposits of the carboniferous, and the backs and arms of the United Mine Workers of America.

All gone, with a twist of the switch.

rules
146. "Don't play with it" (Chandler, kAPP, tPRO).
147. "Keep fingers off" (Denis, tAPP).
148. "Don't breathe on bottom" (Randall, tAPP).
149. "Don't tilt the shade" (Randall, ktAPP).
150. "Don't take the shade off and use it as a space helmet" (Randall, kAPP).
151. The lamp can be moved "but you have to put it back where it goes" (Denis, ktCON).
152. "Don't leave it on all night" (Randall, Chandler, tCON).

What is the bottom of the lamp that one should not even breathe on it?

It is a precious metal.

This is the meaning of the chrome: that the lamp is a Lamp (it is made of silver, it is made of platinum). Here the threshold of significance has been raised to ridiculous heights, not only because the chrome is so easy to sully (and then has to be polished: Wright's Silver Cream "is ideal for cleaning Chrome, Stainless Steel and Porcelain"), but because this badge of the shiny and the bright . . . *is at the level of the feet.*

Actually, the chrome is no more than a relay. It is the poor man's stainless steel (which is Modern Man's "timely metal"). And in fact the lamp (and that by the side window [42]) was purchased in reactionary defiance at a time of real poverty (the lamps took the rent money, which Denis then had to borrow from his brother, Christopher). The lamp thus represents not only the anamnesis of an economy but what made it possible, namely, that recuperation by the bourgeoisie of the classicizing elegance (the lamp appears as no more than . . . base . . . neck . . . shade) and revolutionary romanticism (the exposed metal) that was the inevitable fate of furniture designed on modern principles. The tell-tale tokens? Chrome instead of stainless steel (chrome pits, it corrodes). Linen instead of plastic, instead of mulberry paper (the traditional material for shades tempers the gesture of exposing the metal [it is a tease instead of a strip {the g-string stays on}]). Finally, the way the base of a truncated sphere draws attention to itself (here is Achille Castiglioni on lamp design: "Any device made for lighting can be considered a proper exercise of industrial design only if the fixture-object is subordinated to the effects of light it produces").

The lamp: passes for modern.

The rules reflect this. That so many of them fall into the appearance code (*keep fingers off* is also to protect the chrome) points to its role as decoration. And as if to illuminate the way routines fail to be positively modeled in the rules, important control functions simply make no appearance here. Thus, we already know when the days grow short that the lamps function as a sign of the family's presence (and implicitly its receptiveness to visitors [it is for this reason that people put their lights on timers when they go on vacation]). Yet no rule was collected to say *when it gets dark outside you cut on the light.* Similarly, *don't leave it on all night* scarcely acknowledges the choreography of switching that moves the light around the room. For instance, the lamp by the couch is turned off during dinner to not obliterate the subtler pleasures of candlelight, but that by the window (42) is turned on to not leave the room in darkness (after dinner that by the window is turned off, that by the couch back on, as someone settles beneath it with a book, a stack of papers). *Put it back where it goes* only hints at the travels of the lamp from the couch to wicker rocker to its usual place, itself a compromise between the pragmatics of being able to pass beside the couch (the lamp gets in the way) and the aesthetics of not wanting the lamp to seem nailed to the wall (the luxury of space to spare).

Don't play with it recalls that the kids once knocked the lamp over (the shade tilts, the chromium mirror offers fisheye reflections, the lamp will rock on its base). Denis and Ingrid made them pay to replace the shade *(see what happens when you fool around),* but managed to make do with the old one (there is Scotch tape, it's a little out of shape). This indicates—compromises notwithstanding—that the lamp suffices: a life without space for the futile, the vapid, and the empty has no time for perfection.

No, really!

The Wicker Rocker

Frame of nothing, the spooled and plaited wicker, its braid of reed a braid of air. It is a chair of perforations, a matrix of bubbles. It is built like a tank.

It looks substantial, this armchair on rockers, but it weighs no more than a year-old child. With a rocking volume of more than twenty cubic feet, the chair has the density of whipped cream, less, of a meringue. A Heywood Brothers and Wakefield Company rocker of almost a hundred years, it was handwoven in Massachusetts of reed imported from China. Reed is rattan that has been stripped of its canes. It seems only to have been painted once (unlike rattan—which could only be lacquered—the reed could be painted), and this long ago, in a coffee-bean brown only slightly darker than the aging reed. The hardwood of rockers, legs, seat frame, braces, and corner blocks has also been painted, but the back cushion and seat have been recently recovered by Ingrid in a fine corduroy of a bright yellow-orange.

If you shake the seat you can hear something rattle, like a loose spring, but there's no telling from sitting on it.

rules
153. "Don't play around with this" (Randall, tPRO).
154. "Don't put your feet on this chair" (Chandler, ktAPP).
155. "No standing on the rockers" (Ingrid, ktPRO).
156. "Don't play around with the wicker" (Randall, kAPP, tPRO).
157. "Don't kick it" (Randall, kAPP, tPRO).
158. "Be real careful with this one, because we destroyed the last one" (Randall, kCON, tPRO).

What is so liberating about metal furniture?

At least two things. First, it subverts an orthodoxy: wooden and upholstered furniture. This permits metal to play revolutionary to mahogany and damask's petite bourgeoisie (we have already seen this: it is Charlotte Perriand screaming, "It is a revolution. . . . Brightness—loyalty—liberty"). Second, it unmasks the concealed (this excites because it is forbidden): where metal had been used in furniture, it was disguised (by gimp) or buried (deep in the seats). It is this that especially distinguishes David Rowland's chair (29): he used not merely metal (so did Stam, van der Rohe, Breuer) but a base one heretofore hidden in the lower depths of stuffed chairs. Stripping away the decking, the rubberized hair, the padding, the seat cover, the cushion has for David Rowland the effect of divulging a secret: "A small amount of wire can furnish great strength with very little weight or bulk. It can provide rigidity, but also resiliency. *And wire suggests intriguing visual ideas: its openness, for example, and its literally infinite variety of forms.*"

All this ends in a new orthodoxy (new things will be hidden [welds, joins]). Ingrid rejects this emphatically: "I will not have a living room that looks like something in an architecture magazine." Wicker makes this refusal possible, without at the same time rendering the room *retardataire*. If modern furniture is "artificial"—fiberglass, complicated alloys, elastomers of polyvinyl chloride—wicker is "natural"; if modern furniture is "machine-made," wicker furniture is "made by hand"; if modern furniture little considers comfort (Philip Johnson has said, "I think that comfort is a function of whether you think a chair is good-looking or not"), wicker is comfortable *first of all*. Yet the rocker is not without its modernist allegiances. Like David Rowland's chair, the wicker rocker is . . . *mostly air;* and Thonet takes pains to acknowledge, though "constructed of completely different materials from its bentwood ancestor," that the Sof-Tech chair "is quickly recognizable as a descendent of the Vienna Café chair" which, as bentwood and cane, has evident affiliations with wicker (thus rocker and Sof-Tech chair are like rhymes with shared etymologies). As for the Mies van der

Rohe lounge chair (53), it has a seat and back of saddle leather, but has always been available in a continuous lacquered caning (that is, wicker); and Mies van der Rohe shared with David Rowland an interest in early Thonet bentwood, producing numerous sketches of rockers, bentwood chairs, and even bentwood rockers (thus rocker and lounge chair share etymologies but their rhymes are suppressed . . . by saddle leather).

The wicker that achieves this refusal, however, was not introduced for this purpose. It entered an early incarnation of the room from Ingrid's apartment (it had been given to her when she moved to Worcester by the mother of a college friend), and it was saved from the trash at the time of the move to Raleigh over Denis's objections by Ingrid's insistence *on its comfort* (but Ingrid had always liked wicker [she had grown up with it on the porch]). Only later, against the incursion of high-style furniture (the lounge chair, the Barcelona table), was it *preserved* by the declaration of Ingrid's principles (at the time, Denis sought a "purer" room, that is, one more like that of his parents, the Saarinen coffee table, the Saarinen "Womb" chair, a Mathsson armchair: nothing "old" [it was all designed in the 1940s]).

The plural of the room is thus attributable to "genetics," but it is further enhanced by a practice of the "find." For the wicker rocker is not the wicker armchair Mrs. McConn gave Ingrid (that's in the basement awaiting repair [it's the one Randall refers to when he speaks of destroying "the last one"]). Long after it had become impossible to sit on *(you can sit in that chair if you wish, but let me warn you the bottom is likely to fall out)* it kept its place *to preserve the plural.* One day Denis walked two little kids home who had been playing with Randall and Chandler. On a pile of trash in their backyard he saw a wicker rocker: the little kids' father's father's (it had been in the basement for years). "You want it? Of course you can have it," the kids' father said, and Denis carried it home on the top of his head. The Woods sat on it straight off, dizzy with the unfettering made possible only by such unexpected "finds." The rocker thus entered the select company of the wire on the mantle (63), the collages (32, 33), Peter's box (31), certain records (27), and the Ernst (20) in the delirium of the *trouvaille.*

Rocker: toy, refusal, find.

XXIV · REPRODUCTION

We talk as though reproduction were no more complicated than copying, perhaps even moving the furniture itself, from parents' to children's homes.

Though some aspects of the relationship between Nancy and Jasper's and Denis and Ingrid's living rooms may have promoted this, it is a ridiculous way to think about it; and all we need do is recall Jasper's parents' living room—that is, Alice and Lehman's—or Nancy's parents' living room—that is, Mary and Wray's—to dispel this notion, even without acknowledging that it is not *Denis's* living room, but *Ingrid's* too, and evidently does not reproduce in any direct or simple way the living rooms of both Nancy and Jasper *and* Erich and Gerda (to say nothing of Erich's parents' living room—that is, Eggert and Magda's—or Gerda's parents' living room—that is, Hans and Marie's). In this way we can see that the living room arises from at least two sets of relationships, that between the generations and that between the lineages.

As to the first, it *is* evident that Denis and Ingrid have reproduced in a relatively straightforward way the living room in which Denis matured (the wicker rocker [35] is the notorious exception). If no more than a couple of *things* actually came out of Nancy and Jasper's living room—the couch (30), certain records (27), the speaker in the sewer pipe (but not the sewer pipe [21])—the *room* is nonetheless the same, the rugs on the highly polished hardwood floors, the paintings on the white walls, the modern furniture, the folk art. Too many people say on entering it, "Ah, a Wood living room," for there to be any doubt.

But Nancy and Jasper's living room does not have the same direct and simple relationship to those of their parents. Yet the room they produced, and which Denis and Ingrid reproduced, was in no sense a free invention (no more than the convergences with the very different living room of the Hansens convinces us of this). Nancy and Jasper did not produce a room resembling that of a Tibetan nomad, a Zinacantecan farmer, a Zen monk of Japan; nor even that of a first-generation Hungarian on the West Side of Cleveland, a black from Scovil Avenue, or most of Denis's *Plain Dealer* customers in Cleveland Heights. In fact in everything but *turn of phrase* the living room they produced was that of their parents, rising fractions of the bourgeoisie on either side (thus, Jasper's father was "in advertising," but his grandfathers were a mailman, on his mother's side, and a railroad engineer; Nancy's father was also "in advertising," but her grandfathers were an insurance executive, on her mother's side, and a painting contractor). On either side of the space between the generations, a couch faced a coffee table with an armchair and lesser secondary chairs scattered here and there about the room. Walls were covered, windows curtained. Knickknacks stood on shelves, paintings hung on walls. At the same time it would be pointless to minimize the differences. If his parents' walls were papered, Jasper's were painted. If theirs had a floral pattern, his were white. If their windows were covered by blinds, sheers, valences, and drapes, his had curtains. If theirs were damask, his were sheets. They will have a tufted couch with a crest rail and turned feet. He will have a Paul McCobb with low clean lines. They will have a knickknack shelf. He will have a primitive American hutch. They will display china shepherdesses, miniature flags of North Carolina, Dixie (this in Cleveland). He will display Day of the Dead figurines from Mexico, an Ashanti doll (no less in Cleveland). Everything is different: the room's the same (it is that of Denis and Ingrid [who have merely gotten rid of the curtains altogether]).

One way of thinking about this is as language: same sentence structure, different vocabulary. Only in this way can generational differences be read with equal clarity on both sides of the space between them (the couch is more or less "traditional," more or less expensive). *Re*production here thus refers to every *production* made in dialogic relation to that of a parent. While we believe this to be an important mode of production (even though it does relegate children to reactionary postures: "Dad was right [wrong] about this [that]. I'll do the same [the opposite]"), we do not lay over-much stress on it in general; for far more powerful is the reality that Denis's father grew up in the room of his parents, where in learning the rules he took over the values and meanings embodied in them, which he later reproduced: living room as site of consumption, contemplative arts, conversation (kitchen in the back [it was a railroad apartment: the kitchen was *way* in the back]). Powerful memories for Denis: Jasper and Lehman shouting at each other (women in the background [or kitchen] . . . wringing their hands): *we are at war in Korea—NO it's a police action* or *Indians* [from India] *are Aryans—NO they're Nigras.* Lehman's face is bright red, Jasper shaking with rage: *both respect the furniture.* It is classical living room behavior (later, Denis, Peter, will play Jasper to his Lehman) gyred . . . *out of control:* one or the other family will leave (another powerful memory: Jasper takes Denis and another brother to spend the weekend with their grandparents, a fight ensues between Jasper and Lehman, Jasper and the boys . . . return home). In all of this, in all of the often even physical fights that will take place between Jasper and Denis, Peter, there will be . . . *no feet on the couch.*

And thus (whatever else, Surrealism, anarchism): respect for the culture that makes it possible to manifest the differences at stake. Here is how R. D. Laing and D. G. Cooper (reading Jean-Paul Sartre) have put it:

> The child does not experience his alienation and reification first of all in the course of his own work but in the course of the work of his parents. It was not rent from property nor the intellectual nature of his work that made Flaubert belong to the bourgeoisie, but the fact that he was born into a family that was already bourgeois. *He accepted the roles and gestures imposed upon him at a time when he could not comprehend their meaning.* But, like all families, that of Flaubert was a particular family, and it was in the face of the particular contradictions of this

family that Flaubert served his apprenticeship as a bourgeois. No change was involved. He lived, in its particularity, the conflict between the reviving religious pomp of a monarchist régime and the agnosticism of his father who was a petit-bourgeois child of the revolution.

With all appropriate changes being made, yes . . . *exactly.*

The Throw Pillows

No chair can be right for everybody, or even anybody, all of the time. That's what the throw pillows are for, that place in the small of the back that the cushion on the rocker *just can't get.* There are two of them, a larger one thirteen inches by twelve and a smaller one eleven inches square. Together they're seven, almost eight inches deep. To make them, Ingrid saved rags and the little scraps of cloth she finds so hard to throw away. The cover for the large pillow combines a fine piece of weaving Denis's parents bought in Oaxaca with a back knit from a gray yarn Ingrid just . . . happened to have. The cover for the other she made from a piece of alpaca a neighbor had given her when he had a little thing on the side importing stuff from Peru.

"It was just a small piece of material," Ingrid says, "and this is what I did with it."

rules   159. "No throwing the throw pillows" (Denis, kApp, ktcon, tpro).
160. "No whacking people with them" (Randall, kApp, kcon).
161. "No pillow fights" (Denis, ktcon).
162. "Don't leave them on the floor" (Chandler, kApp, tcon).
163. "Straighten out pillows after use" (Denis, tApp, kcon).

Why not throw the throw pillows?

Because "throw" is hyperbole; it doesn't mean "to propel through the air by a forward motion of the hand and arm," but no more than "anywhere" when the question is "where do they go?" But the hyperbole is neither adventitious nor expendable. It denominates precisely that *luxuriance of the loose* that infects anything that is not tied down. They can know no place because they have no place (rising above it, they constitute a law unto themselves). Actually, *this* is hyperbole, though *don't leave them on the floor* is the only rule to insist upon it.

Compare the couch cushions (30). Loose, still they may not go anywhere. If the back cushions may be piled beneath a head, the seat cushions may not be moved (RULE 141: *don't take up the pillows and sit under them*). Yet even the back cushions have their place (furthermore, this is specified by the shape, size, color). The lamp (34)

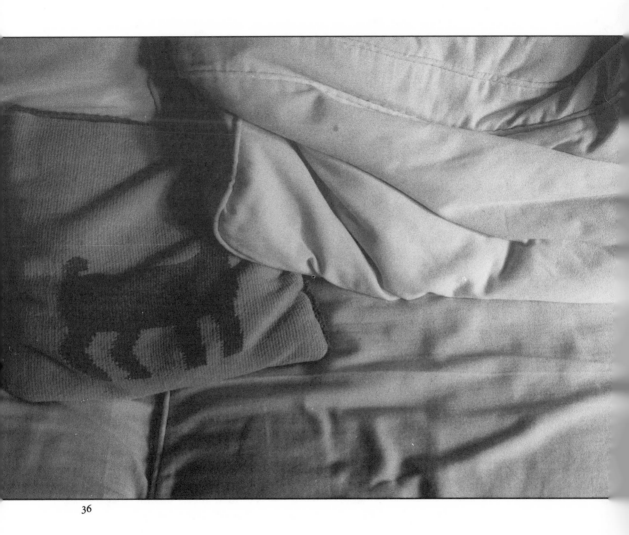

also is loose (it is on a leash), but it too has a place (out of the way of the passage beside the couch, but not nailed against the wall). Even the couch and table (65) are loose, and both are moved, to wax the floor (the table is even disassembled) and at Christmas when they're pulled back out of the way (then the larger throw pillow gets a seasonal cover and the smaller one is replaced by a seasonal pillow). Little is tied down (the floor [8] is nailed down, the newel post [10] is tied to the framing). Against these dignified seasonal currents the motion of the throw pillows is of another order, is that of a froth, of juggling (wonderful that this excess embodies the parsimony of Ingrid's rag bag [like beanbags]).

Herein, of course, their utility: to be where the back (neck, head) needs them. For hardships: luxuries.

Ron's Drawing

An extraordinary moment: to come from the kitchen and find, ruddled by a late fall sun, the Ferris wheel within the frame of Ron Rozzelle's *Encounter: The Fish and I*.

This is a drawing, in colored pencils on paper, of Ron talking to a fish. The wooden frame, a little more than twenty-eight inches by thirty-five, is glazed with Lucite.

hand-off   2. "Same as every painting" (Chandler, Randall, to RULES 90, 91, 145).

Denis has written about this drawing before. In fact, it was while writing the catalogue essay for a show of Ron's drawings at the North Carolina Museum of Art that Ingrid asked him, "So which one are you buying?" Until that moment he had not realized that he was.

There were four drawings in the show, all "encounters" with animals: a wolf, a hawk, a fish, a snake. In response to a rhetorical "What are these all about?" Denis wrote:

> One thing they're *not* about is human encounters with wildlife. Had all four drawings shown Rozzelle fending off attacks from raptorial beasts, then maybe it would have been possible to imagine them as covers of a sort of high-class *Argosy* magazine ("I Battled a Wolf !!," "Attacked by a Hawk!!"). But they don't do this. They show him holding a conversation with a fish and engaging in diplomatic negotiations with a snake, though even had *The Snake and I* shown Rozzelle escaping from the python's coils, something still would have been missing: the Banana Republic clothing, the jungle foliage, the gun lying useless on the path (and all the rest of the details so important to the *Argosy* tone). The thing is,

Rozzelle didn't just omit these details. When you think about it, you realize there is no setting in which both Rozzelle and the fish could actually breathe together, that *The Fish and I* exists in a world *Argosy* has never explored, not just out of the time and space of our ordinary life, but outside time and space altogether, that these are not drawings of encounters with animals at all, that they are something else.

What they are, are symbolic dramas: dramas on the face of it, and since the encounters cannot be taken literally, clearly symbolic as well. *The Wolf and I* is not, thus, a drawing of Rozzelle beating off *a* wolf (*any* wolf) that just happened to wander up to his tent, but a drawing of Rozzelle taking aim at *the* wolf, his own *personal* wolf, the one that in his imagination (in his dreams) has come to stand for . . . what? He doesn't say: the fear of hunger? the temptations of fierceness and cunning and greed? The Middle Ages made of the wolf a sign for heresy; Jack London turned him into a symbol for all that is wild. For violinists, the wolf is a sour tone lodged in the belly of their instrument. And for Rozzelle?

But does it matter what Rozzelle's wolf is? The wolf is at him whatever it is, as yours must be at you, whatever yours is, the fear of alcoholism, of gluttony, of becoming a lecher: what is it? And how does it feel when it comes on you in the night, fangs bared, ready to leap? What do you do? How do you fend it off, your personal demon? This is what these are, of course, *demons*, or better yet, *daemons* (which takes us beyond the simple sense of evil evoked by "demon"), that is, familiars, geniuses, guardian spirits, attendant ministering powers, daimonions, perhaps even *muses;* or in a language we find more comfortable today, *feelings, urges, luck, instincts, talents, intelligence,* but reified, hypostatized, as primitive peoples do, as the Greeks and Romans did, as shamans and dreamers and little kids do (—as you do—), into godlings and godlets and godkins and goddesses, into sprites and spirits, into Luck and Intelligence, into . . . wolves and eels and snakes and fish and hawks.

It is a drawing, in other words, of a dream, and it reminds us that the rational (fish in water, man in air) is merely one system among many. Having opposed the hegemony of the modern, with the Rozzelle the room takes a stand (at no little cost) against that of the rational (putative ideology of the modern). It also makes clear that the dream need not be a matter of smokes and mists (nothing of Moreau, of Victorian apparitions) and so opens the door in a room of open space, white walls and light . . . *to the night.*

XXV · HOW MUCH?

How much?

An inexorable question (moreover a vulgar one) which nonetheless hangs in any air polluted by the competitive energies of industrial capitalism.

Hangs in the air: this is the key, for the question can be neither asked (vulgar, probing) nor avoided (curiosity, value, meaning), a conflict resulting in a genuine baroque of stratagems for doing so, from the *blurted* (not afraid to transgress the conventions that forbid asking) to the expression of the wish *not to know* ("it would only interfere with my appreciation of the work"). Nor is the owner passive: he may anticipate ("and you'll *never* guess what we paid for it") or evade ("I don't remember"), as counters are exchanged in this game of establishing taste, sensitivity, social status, and worth—that is, meaning and value (the very things turned into rules).

Though the relevance of cost to the meaning and value of things cannot be denied (we do not deny that it *is* denied, often and loudly, only that these denials are anything other than a way of coping with the anxiety raised by the means of producing both thing *and* value), no measure of cost is capable of *exhausting* meaning and value. This is easy to see in no more than the difference between cost and worth (flea markets thrive on the energy this difference releases), where acquisition at a fraction of the worth is often the whole point of display (the owner is valorized as hunter, as connoisseur). Contrariwise, acquisition at a multiple of the value can testify to the purity or strength of an owner's obsessions, attributes which can validate the most "individual" of tastes ("it didn't matter what it cost: I *had* to have it"). Many things in the room of significant worth cost Denis and Ingrid nothing. For instance, Jasper gave them the Paul Strand (56), and as we know they found the wicker rocker (35). Other things of some worth cost very little. For instance, the Ferris wheel cost $25 (it cost the same to ship it), at a time when Alexander Girard's Textiles and Objects was selling it in New York for $250. In some cases, parity reigns: both the Ernst (20) and the wire on the mantel (63) are worth what they cost. On the other side, there are things whose cost exceeds their current worth (the Palaset cubes [23, 26]) and things that whatever they cost are worth nothing (certain records [for instance, the Shaun Cassidys]). Nevertheless, Denis and Ingrid paid for them (often one or the other, sometimes both) and any understanding of the room requires some assessment of their cost (not their worth, which is no more than their cost to some other who might wish to buy them [this could, however, become a measure of their *adhesion* to the room {though confounded by the problem of "saving for a rainy day"}]). At the very least, it will give us insight into the room as . . . *a hierarchy of costs.*

Even this is easier said than done. To control for inflation, we have calculated costs as percentages of Denis's salary at the time of purchase, his salary increasing in small legislatively mandated increments highly responsive to cost-of-living indices. We have ignored all financing costs (in our case, only the cost of the "room" itself was financed) and non-salary income (the question here is not *how did you pay for it?* but *what did it cost?*). Here, then, is everything that cost more than 1 percent of Denis's salary in the year of its purchase:

The house   180%
Ron's drawing   9%
The Ernst   8%
The Mies van der Rohe furniture   5%
The couch   4%
The speakers (pair)   3%
Records (per year)   3%
The Cherokee basket   1%

A list such as this constitutes a class signature as surely as the spectrum of black and white lines identifies an element in the light of a star: it is that dominated fraction ("artists and intellectuals") of the dominant class ("the bourgeoisie") whose position is a function of its possession of cultural capital. Here there is evidence of an almost defiant investment of marginal economic capital in "cultural" goods, one moreover understated, since the cost of the entire house is opposed to the cost of the "contents" of no more than *one* of its "rooms." Dividing the cost of the house by the number of "rooms" makes the "real estate" component a more reasonable fraction of Denis's salary (28 percent), but even this ignores the reality that over time increasing quantities of economic capital are accumulated at higher rates in the things *in* the "room" than at that at which the "room" itself appreciates. Particularly egregious is the way in which at somewhere between 50 and 100 records a year the total cost of the record collection approximates that of the "room": today the replacement cost of the collection exceeds insurance estimates for the replacement cost of the "room" (that is, the replacement cost of the house divided by the number of "rooms"). Our designation of the class fraction ("artists and intellectuals") is confirmed by the eccentricity of refusals that makes the investment in "culture" feasible: no car (no automobile insurance), no TV, no central air, no beach house, no stocks or bonds, no savings.

One of the things validated by this analysis is our reading of the cracks in the wall (16). They are not those of a tenement in a Raymond Chandler novel (sign of a cultural decline): no question of capital, but its investment (mind first). It *is* student housing.

No, really!

Both Denis and Ingrid independently agreed they would save Peter's portrait of Genovevo de la O (64). Ingrid's response was immediate and emotional ("I love that painting"), wrapped up in her recollections of how and when Peter gave it to them ("it was a lot the way he did it, you know, that way Peter has") but not free of a love for its forms and colors. Denis's response was no less immediate, but justified by an analysis of replaceability (one would save unique things, Peter's painting, Ron's drawing [37]) and ideology (that of revolution over that of the dream [since the dream, too, is a social construct]), though it too was not free of a love of its physical presence. Cost, and its correlate worth, entered into neither response (Peter's painting cost nothing, is worth little), which is not to claim independence of value and meaning from cost and worth, but to insist on their complications.

Randall nominated the records (27), Chandler a single cube of records (23), their friend Frank (sitting on the porch swing next to his boom box) the stereo.

"We don't have a stereo in that room," said Chandler.

"Oh, yeah," said Frank.

The Left Speaker

"Beep-beep-beep-beep-beep-PH-E-E-E-T—"

"There's another ship coming in."

"Look at him. He's heading for that small moon."

"I think I can get him before he gets there—he's about in range."

"That's no moon . . . it's a space station."

"Why are we still moving toward it?"

"We're caught in a tractor beam and it's pulling us in."

"There's gotta be something you can do—"

"There's nothing I can do about, kid! I'm full power. I'm going to have to shut down. They're not going to get me without a fight—"

Though it makes more sense *in the room* where the speakers project it stereophonically, giving auditory substance (and distinction) to Obi-Wan and Luke, Han and Chewie, mapping the space of the *Star Wars* universe onto the space of the room, so that when that TIE fighter comes in ("There's another ship coming in") . . . it comes *IN*.

The speakers are locally made—not just assembled—by Raleigh's Creative Acous-

tics, who have mounted a ten-inch woofer, four-inch midrange, and Philips tweeter in a ported enclosure occupying a volume of just under three cubic feet (at sixty pounds this gives each speaker the density of charcoal, surprising, given their appearance, until it is recalled that they too are mostly air). The drivers are hidden behind an acoustically transparent grill cloth stapled to a frame that snaps into place. The cabinets are wrapped in walnut veneer that (ingeniously) runs up one side, across the top, and down the other.

They are powered by a Hafler amp in Denis's study, to which they are connected by cables racing beneath the floor. The Hafler preamp accepts inputs from either of two Nakamichi tape decks, an ADS tuner, and an old Kenwood turntable. To play a record, you find it in one of the record cabinets (*The Story of Star Wars* is under S in the popular section of the small record cabinets [23]), put it on the turntable in Denis's study, and return to the living room to listen to it. Chandler usually sat on the couch (30) in the days when he made this record the most frequently played in the history of the room.

With the album open on his lap . . .

Waiting for his brother to come back from preschool.

rules
164. "Be careful" (Denis, tpro).
165. "Please don't move them" (Denis, tcon).
166. "But we move them at Christmas" (Chandler, tcon, tpro).
167. "Don't unplug wires or fiddle with knob on back" (Chandler, tcon, tpro).
168. "Don't take off covers" (Chandler, tpro).
169. "Don't touch the speakers under the cover if you take them off" (Chandler, tpro).
170. "Don't leave things on it" (Chandler, ktapp).
171. "Don't play the records too loud: if you do, make sure Bill's window is shut" (Denis, ktapp, ktcon, tpro).
172. "Unless Den's working, then we can't play it too loud" (Chandler, ktcon).

What are the speakers? They are the device in the room responsible for transforming the quarter ton of vinyl chloride in the record cabinets . . . into words, into music. Let us recall that these records occupy twenty-five cubic feet of obsessively ordered space, that they are what Randall and Chandler would save from fire, that they are those things in which Denis and Ingrid have invested to the extent that their replacement value today exceeds that of the "room." The speakers make it possible for all . . . *to be heard.*

The speakers themselves represent no inconsiderable investment (they are the fifth most costly thing in the "room"). Furthermore, they are no more than part of a larger investment, which is given voice by them as they are driven by it. Nor is this exclusively technical (turntable, amp, preamp, tuner, decks); it is also ritual (for example, in both the playing and exchanging of music at Christmas); emotional (much of the music is loved); intellectual (hence in Denis's study *Grove's Dictionary of Music and Musicians*); and corporeal (the Woods bop to the music, snap their fingers, sing, whistle, hum, tap their feet, hoot and holler, Ingrid and Denis dance).

But even at the level of equipment, the speakers are not exhausted by their phonic function. They are also furniture. As such they are modern (their lines are clean and simple), but distinctive. They introduce into the room both black (which paradoxically does not oppose white but joins it to enrich the modernist palette [the two colors unite in their extremist purity to oppose the striped, the spotted, the checkered, the plaid]) and finished wood (otherwise relegated to the floor [8] except for the fugitive echo in the frame around Ron's drawing [37]). Here we are enabled to understand that the apotheosis of metal does not imply an abjuration of wood (nonetheless: no mahogany, no Brazilian rosewood). Furthermore the speakers are also pedestals for sculpture. As such they constitute privileged sites, capable of investing whatever stands on them with increased value (they speak in the exploitative). As we know, this effect is heightened by any subversion of appropriateness, as here where the crude low-fired pottery of the Lacandon drum (39) is placed in tension with the high-tech of the speaker (aluminum castings, powerful magnets, sophisticated cross-over networks); or on the right speaker where the encrusted surface of the earthenware king (55) is set against the black of the grill, the slick surface of the walnut veneer.

Given all this, what is surprising is not the number of rules, but the inaudibility of the appearance code. The excision of this voice marks a concession to a pragmatic which, like that of the stair (10, 11), will *work*. However, unlike the stair, which worked by not playing, the speakers will work precisely *by playing*. Hence where the rules carpeting the stair protected the kids (from falling, from cracking their skulls), these veneering the speaker will protect the speaker, Denis . . . the neighbors. As if savoring the distinction, the rules themselves will play: *don't play the records too loud,* well, since you will, *make sure Bill's window* [50] *is shut; don't take off the covers,* but since you will be unable to resist, *don't touch the speakers* beneath it. What is expressed here is the function of the rule stripped of its normative pretensions. Denis would as soon the kids didn't touch; but the speakers are powerfully seductive, the kids *will* fall, and since they will, an inner rule: *don't touch the speakers.* This could well have been repeated: *don't touch the speakers* (that is, the drivers, the cones), *but if you must* (who can resist? the surface is alive), *don't hurt them* (compare, the speaker in the sewer pipe [21]). This is what really is at stake: that the drivers not be disabled. In the effort to insure this state, the rules establish increasingly weak fields at increasingly great distances from their actual object (like electron shells, like outer defenses, like picket lines) in the anticipation that they will be breached. Pragmatically the rule admits the attractiveness of the pulsing cone, but since the transgression of any rule absorbs energy, by implacing concentric zones of protection, the rule system nonetheless manages to protect it. Another instance: *don't play the records too loud.* But who can resist? Therefore a pericardium, *make sure Bill's window is shut,* around the heart of the matter: *don't bother Bill* (synecdoche for Bill and Kathy, next-door neighbors).

At first *please don't move them—but we move them at Christmas* seems another example, but it's not: *don't move them* is the irreducible nut (at stake is stereo imaging). Here then the contradiction acknowledges not a loss of innocence, but the changes in the rule structure necessary to accommodate the changes in the room attendant upon the periodic celebration of Christmas. What takes place is a kind of phase-shift or a change in key (to pick up Erving Goffman's musical analogy) whereby, for the duration, *everything is altered* (why not? for the duration a profane room has become . . . sacred space). The contradiction thus acknowledges two conditions of (mutually exclusive) work: with and without the Christmas tree (when the tree stands in its corner, the left speaker turns its back to the branches to protect the drivers).

In whatever the key, the speaker will work . . . *by playing.*

The Lacandon Drum

From the darkest corner of the room stares the face of a god.

It is dominated by an enormous nose, swollen protuberant eyes beneath bushy eyebrows and a wrinkled forehead. Below bulbous ears, bulging cheeks frame puckered lips. For all of this, its appearance is wistful.

Its neck merges with the wall of a small bowl, cup, censor, or brazier which merges in turn with that of a vase seventeen inches high, twice that around. The whole is modeled of clay that has been allowed to dry in the sun before being baked in what amounts to an open fire. Though most Lacandon pottery is decorated in red and black on a white background, the evidence here is that it was painted in red and black and overpainted in white. In any case, the decoration consists of a number of (all but obscured) black circles on the vase, and seven stripes on the censor below the face. Two pieces of vine—or perhaps they were branches of a tree or shrub—have been lashed into rings around the base and neck. They are held in place by a network of twine— apparently store-bought—that girdles the vase. A piece of deer skin secured to a third

ring of wood or vine has been stretched across the opening of the vase, lashed in turn to the ring around the neck. The drum would have been played with the fingers, while *copal* (or *pom*) burned in the censor and prayers rose in the night to Hachäkyum.

It would not be quite correct to think of this as a representation of *k'ayum*, the god of music, so much as the god itself. *K'ayum*, from *kay*, to sing; and *you*, gentleman, lord, master: *he who sings.* Who with more justice could stand . . . *on a speaker?*

rules   173. "Under no circumstances touch this drum" (Denis, tPRO).

174. "Don't play with it" (Randall, tPRO).

175. "Don't put marbles or anything in the cup part" (Chandler, tCON, tPRO).

We know that the interdiction of touching is the sign of the precious, but this crude pottery drum is not lace, is not a gem. What is it?

Inescapably, it is a trophy.

That is, it is the memorial of a victory, a conquest (it is the spoils of a hunt [it is the souvenir of a tourist]). This would be the case even if the Woods beat the drum, burned *copal* in the censor, and venerated the Lacandon gods. But they don't. Removed form the matrices of oppositions and affinities in which it was created—those of the Lacandon world, everything that would pass for its cosmology, its ritual practice—it has been stripped of everything that made it *k'ayum*. Prevented from even being touched, it is no longer *he who speaks* but *that which is wondered at* (it is an ethnographic object [that is, a freak in a carny show]).

It would be hard to overstate this quality of the drum. Armed with money from his paper route (and a favorable rate of exchange), in Mexico Denis *hunted* with his family for the things that he bought. Led by guidebooks (Kate Simon, John Wilhelm) to the most likely stores, he and his mother also stalked the streets, entering each shop—however unlikely—with an *"¿Hay juguetes?"* A few clay whistles? *He would take a dozen.* A handful of papier-mâché masks? *He would buy them all.* Finally he was admitted to the storeroom secreted behind a shelf in the store run by Joaquin Hernanz Humbrias, where in a paroxysm of consumerist euphoria he was permitted to see and later buy the drum. It was carried to the hotel in what it would be disingenuous to call anything but a triumph as, back in Cleveland, were the spoils unpacked, the display of wealth brought back from . . . *the New World,* among which the drum was a special treasure, the artifact of a dying people (there were said at the time to survive between 300 and 400 Lacandones) the more precious for its fragility, for the unlikelihood of its survival of the trip to Ohio (whose own name is Indian [it is Iroquoian for *beautiful*]).

Is the drum exhausted by its presence as imperialist plunder?

Hardly. For this could as well have consisted of duty-free tequila, sombreros with *Mexico* stitched in red, tee-shirts proclaiming *Acapulco.* It did not. This drum (not an onyx chess set), ripped from the matrix of oppositions and affinities that gave it meaning, acquired new meanings in the matrix of oppositions and affinities into which it was plunged. Except insofar as they subsist in its form (it is white, it glows in its dark corner), these are inevitably parasitic (there is no escaping the imperialist sting)—that is, they consume the drum as *a sign function* (what it was to the Lacandon is of the essence [for certainly it is not a drum in the room {*don't touch*}]). Thus, though not a god for the Woods, that it was for the Lacandones *matters*, and in at least two ways.

First, it is precisely this focus of attention that makes the drum an ethnographic object (not an objet d'art, not a musical instrument) and which permits it to slip into that vast slough of attitudinizing that has characterized European images of the New World from the beginning; themselves deliquescing into nostalgic images of a classical West (arcadian Greece) and exotic images of a fantastic East (spices, silks, fabulous riches). Inevitably these infect with an unsuspected virulence any posture one adopts toward any Indian artifact, but especially here where the recent intrusion of large-scale logging in the Selva Lacandona replays the arrival in the New World of Europeans in every particular: theft of land, destruction of habitat, importation of disease, decimation of population (to such small numbers that it is no longer possible to speak of a Lacandon people). Here the drum is wrapped up in notions of the primitive, of innocence, of purity and simplicity, of freedom from the intrusions of private property, that at once connect it to Peter's box (31) and his paintings (41, 64)—as well as through them to a generalized populism—and set it in opposition to, among others, the speaker it stands on, thereby joining the Cherokee basket (24) in its saturation of the paradigm *primitive/sophisticated*. Most notably evaded in this way is the progressivist image of a history that chronicles, above all else . . . *the ascent of man* (that is, double irony, his escape . . . *from the jungle*).

The second way its being a god for the Lacandones matters is to see in the drum that key to that *enchanters' domain* the surrealists found in sculpture of the Americas and the Pacific islands. After observing that "painters and poets were among the first ones to jump over the fence on which the specialist had always kept close guard, intent as he was on preserving from sight the objects that belong to the paraphernalia of ethnology, and that would henceforth pertain to the 'primitive arts,'" Vincent Bounoure goes on to speak in these tones about surrealism and the savage heart:

> From one shore of the Pacific to the other, from the Indian Archipelago to Araucania, including Melanesia, Australia, British Columbia and the Pueblo Indians, the same form of apprehension of the world manifests itself: a universe as impassioned as can be, usually composed of two complementary halves, a uni-

verse conceived as an androgyn, not a primordial androgyn as in the speculations of antique Mediterranean civilization, but a still present and active androgyn, every essential gesture of whom is a love drama. The state of interpenetration of the world in which man lives and of his personal idea of the world is such that this very pattern is apparent in the organization of society; sometimes it is more geometric, as is the case among the Navajo Indians whose sand designs are the compass-dial of a sacred space, and sometimes more openly sexualized, as in Australia or Melanasia. Nothing less than that was needed to color with the hues of the marvelous the works which so manifestly expressed this high state of conscience. Perhaps surrealism would not have become what it is had it not developed in the shade of the malangas through whose branches the night breeze stole among networks of lianas and snakes.

Again the jungle (apparently an inescapable term); but though it may be that of Henri Rousseau, nevertheless it is neither that cradle of innocence (Dryden: "I am as free as nature first made man / Ere the base laws of servitude began / When wild in woods the noble savage ran"), nor that site of savagery (law of tooth and claw) evoked by the opposition of primitive and sophisticated; but *something else altogether* whose articulation is first of all an evasion of precisely these terms of discourse (Bounoure again: "an altogether different dimension of the mind which had madly ventured into the jungles of dreams where scarlet birds rule").

These new terms? *Rational thought/pure psychic automatism, technical use/critical-paranoiac activities, art as language/trajectory of dreams.* Swiftly they create a space where Ernst's bird (20) and Rozzelle's fish (37) dance to the beat of a Lacandon drum.

It makes of the room a dream land, it pushes it (a little) . . . *over the edge.*

Adja's Lithograph

On the white wall above the drum, behind a sheet of plastic that scatters here and there the ghostly images of lamp and window, in blue, black and yellow . . . a medallion for the room.

It's *L'Amoureux,* lithograph in three colors by Adja Yunkers. The medallion measures thirteen inches by just under eighteen, but it was printed on a sheet of Arches twenty-two and a half by thirty. According to Jo Miller's catalogue entry in the Brooklyn Museum monograph on Yunkers' prints, *L'Amoureux* was issued in an edition of twenty-five, but the one in the room is hors concours, III/V, and signed and dated (1966) in the lower right.

The frame, a Kulicke Trap XM, is pure Lucite, The intention here is to . . . disappear.

hand-off 2. "Same as every painting" (Chandler, Randall, to RULES 90, 91, 145).

To say that the Yunkers is not immune to the currents of meaning flowing through the circuits linking Ernst (20), Rozzelle (37), and Lacandon drum (39) would be no more than to acknowledge the aptness here of the phrase "abstract surrealism," one suggested by Robert Motherwell to designate that school of lyrical abstraction that flowered in New York under the sun of a surrealist automatism (Masson, Ernst, Matta). Without doubt Yunkers accepted without cavil his frequent description as an "abstract expressionist," yet Motherwell's distinction remains nonetheless real; for even among those most profoundly influenced by the surrealists (Baziotes, de Kooning, Motherwell, Pollack, Rothko) Yunkers—perhaps by virtue of his superior age, perhaps due to his European background—was especially so, to the very end of his life. The Yunkers thus contests in at least three ways the syntagm of classical patriarchal Western rationalism installed in the room through the competence that constructed door and wall, stair and floor: first, through the implicit surrealist attack on the bourgeois habits of rationalist thought (/ *pure psychic automatism*) and technical utilitarianism (/ *critical-paranoiac practices*); and, second, through the explicit assault abstractionism makes on art as language. This, which has proven fatally easy to recuperate (what looks as good in a corporate office as an abstract painting?), is the potentially more subversive element, for it constitutes as well an attack on language as communication, or, at any rate, language *reduced* to communication (and hence, unavoidably, Adja's *expressionist* program). A bonus: the intrusion of abstract imagery into a room dominated by figurative art (20, 32, 33, 37, 41, 56, 64) constitutes a further contestation for it; and not only absolves the room from having to debate the relative merits of abstraction and figuration—and hence preserves it from the hegemony of either—but thereby aligns itself (once again) with the surrealists, who, it must be admitted, dismiss both parties as unsuited (only painting that speaks . . . *with the voice of liberty*).

Do Chandler and Randall understand any of this? Probably not in this way, but the charge that images such as *L'Amoreaux* cannot be read is one that is vacated by its own history, one moreover which not only surfaces the contradictions in bourgeois thought regarding abstract art but impales the room on them. Denis's parents bought their first Yunkers a couple of years after they bought the couch—that is, in that period of transition following the move to Cleveland Heights from the housing project. This led to an acquaintance with the artist, which flourished after the Malmquist and Wood Gallery began to represent him (one-man shows in 1964, '66, '67).

Given that Malmquist and Wood was primarily an advertising art studio, it is less than a surprise that one of its clients (the advertising agency, Griswold and Eshleman) commissioned Adja to make a print for one of *its* clients (the Addressograph-Multigraph Corporation) as the inaugural gesture in a campaign to publicize a new machine. In the event, the responsible executive at Addressograph-Multigraph proved unable to countenance the representation of his company by an obscene image, one, in his perception, of a vulva. Here we see first, in the commission, the recuperation of *abstract art* in the name of a certain prestige (today's equivalent: corporate sponsorship of modern dance). Yet by virtue of a miraculous insight (that of Mrs. Grundy) this abstraction is revealed as figurative, furthermore, as a *masked* figure (wolf in sheep's clothing [smut in the guise of art]). Therefore: its rejection as *figurative art* in the name of a certain morality. For Adja, of course, another issue entirely: only painting that speaks . . . *with the voice of liberty* (yet to make a living in a world where only the wealthy buy art [that is, those most certain to be strenuously offended by any whiff of subversion]). The result? Griswold and Eshleman eat the edition, which is distributed "in house" as a kind of perk.

Therefore, two points: one, Randall and Chandler *will see* the lithograph, doubtless with different eyes in every succeeding year (and certainly they have heard Denis tell this story); and two, it is *this story* no less than the image which will be internalized: art, corruption, betrayal, a hint of sex, the confusions of kept birds . . . *on the wing.*

XXVII · THE DOMINATED FRACTION OF THE DOMINANT CLASS

In the history of the lithograph (40) we see very clearly the antagonism between the dominated fraction of the dominant class ("artists and intellectuals": Adja, Denis's father [Denis]) and the dominant fraction of the dominant class (the "bourgeoisie": the responsible executive at Addressograph-Multigraph). This antagonism—and its rhetorical embellishments (*perverse obscenity/ Mrs. Grundy*)—obscures the essential commonality of the relationship of these two class fractions to . . . *necessity;* and therefore to the dominated classes perpetually subject to its strictures (the "lower classes"; the oxymoronic "subclass"). Little more than the suspension of the room in this context is required to "explain" the major currents of meaning coursing through it—that is, to reveal them as none other than manifestations *in the room* of the social relations obtaining outside it.

The major contestations the room makes in this way appear as embodiments of the antagonism between these rival fractions of the dominant class (that the room is not that of the lower classes goes without saying in the same way that the room's orthogonal planes pass . . . unnoticed [how else could you build a room?]). For example, it is characteristic of the bourgeoisie to immerse themselves in a world of social denial (a taste for landscapes, especially those of the Impressionists; or if for people, those of Renoir, Mary Cassatt; period furniture; drapes; Brahms, Mahler; comedies of manners). Therefore *the room* will overflow with social awareness. If it can imagine three kinds of society (*"primitive"/ peasant/ industrial*) it will allude to each (*the Lacandon drum* [39]/ *the Ferris wheel* [25]/ *David Rowland's chair* [29]). If it understands three classes (*folk/ popular/ elite*) it will instantiate them all (*the throw pillows* [36]/ *the wicker rocker* [35]/ *Ron's drawing* [37]). If it recognizes national differences (*France/ the Qualla Boundary/ Finland/ Belgium/ England/ Peru/ the United States/ Mexico*), it will acknowledge as many as possible (*the Ernst* [20]/ *the Cherokee basket* [24]/ *the record cabinets* [23, 26]/ *the French language* Tintins [28]/ *the British* Tintins [28]/ *the fabric on one of the throw pillows* [36]/ *the doorframe* [3]/ *the wooden car* [60]), emphasizing not only countries highly alien to an Anglo-American bourgeoisie (Mexico, Peru), but within any country those elements most antagonistic to it (in Mexico, Lacandones, Zapotecs; in France, the surrealists; from England, translations of Belgium comic books). Or again, there is a bourgeois taste for the decorations and ornaments of a certain ostentation. The room will therefore refuse in the name of an ascetic purity every superfluity (no drapes, floral wallpapers, brocaded upholsteries: only the essential [a corresponding taste for the Bach of *The Well-Tempered Clavier* {but on the harpsichord, not the bourgeois piano (where a taste for . . . Satie)}]). In general, the room will reject the facile, a respect for facility, the showy (a repudiation of the virtuoso performer, a lack of interest in the concerto), but will embrace the popular (rock and roll, the Clash, Shaun Cassidy [épater le bourgeois!]). Or again, it will display tokens of an explicit struggle against the bourgeoisie: surrealist artifacts, portraits of popular revolutionary, anarchist heroes. That the room defines itself in opposition to the bourgeoisie cannot, however, be permitted to obscure the fundamental reality that the ability to do so is grounded in the shared relationship these rival fractions have vis-à-vis the dominated classes. Thus, it cannot be overlooked that the organization of the home is thoroughly bourgeois: in a two-story single-family house the opposition of living room and kitchen, shared meals in a dining room, separate bedrooms. More critically, the ability to saturate a paradigm like *folk/ popular/ elite* depends on the dual luxuries of having, first, the economic capital to instantiate the elite term (the wherewithal to pay for the Mies van der Rohe table) and, second, the cultural capital to instantiate the popular and folk terms (the self-assurance to confront the table with a wicker rocker).

Does this mean that Denis and Ingrid bought the Mies van der Rohe table to instantiate a term in a paradigm?

Not at all. They bought the table because they loved it, its cool elegance, its Mack truck–like construction, its utopian promise of reconciling an aesthetics of purity and refinement with the social realities of industrial production (if they were filling a term in a paradigm it was that of *kitchen table/ dining room table/ living room table*). On the other hand, (a) their love of the table was not "pure"—that is, it was not free of the formative influence of the class in which they matured (rising fractions of the bourgeoisie investing heavily in education), and (b) we know that Ingrid insisted on the wicker to *violate* the syntagm of "a living room that looks like something in an architecture magazine"—that is, to activate an opposition in the paradigm *like an architecture magazine/ not like an architecture magazine* (this in turn is not pure: Ingrid had always proclaimed the value of wicker's comfort; furthermore, she grew up with it). *But this mixture of motivations is precisely the experience of living the dominated fraction of the dominant class, which inevitably loves those things that allow its class fraction to express itself in its living.*

In this sense the room is an embodied sign of Denis and Ingrid's class solidarity (it may attack the bourgeoisie, but it also stands in thoroughgoing opposition to the dominated classes to which, in any event, it does not open its arms), and what is reproduced in the kids through the rules that embody its values is . . . this class fraction.

Is it possible to stand outside class?

We don't see how.

Does this mean that this book is written from within the perspective of this dominated fraction of the dominant class?

Yes.

Peter's *Malatesta*

Confusing miracle: an absence of glazing increases this painting's resemblance to a window; yet, instead of absorbing light, it seems to transmit it, to another world.

It is Peter's painting of Malatesta, made in 1970 with oils on board using nothing but his fingers and hands (no brushes). At better than thirty inches by forty, it's not small, though Denis's brother Christopher has a later version that is six times the size.

It's in the classic Kulicke museum frame, a slip of aluminum, welded at the corners, then buffed to a dull sheen. The aim here? Discretion.

hand-off 2. "Same as every painting" (Chandler, Randall, to RULES 90, 91, 145).

Malatesta is an important painting, the touchstone in the room of that spirit of anarchism that touches—if only with its wings—so much of the living that is the family. During the writing of this book, Denis's father, Jasper, called: "Tell them I'm an anarchist, too"—he whose living room bristled with more rules than the room we're studying, and this despite a surfeit of Kropotkin and Read, Woodcock and Comfort. But no more than Kropotkin's *Britannica* definition—"Anarchism, the name given the principle or theory of life and conduct under which society is conceived without government—harmony in such a society being obtained, not by submission to law, or obedience to any authority, but by free agreement"—was needed to make it clear to his adolescent sons that whatever Jasper meant by his claim, his home would not be characterized by a harmony obtained through free agreement. The converse, in fact, bitter discord—mimic of every internecine battle of the Left—sustained through the endless renegotiation of ultimatums delivered ex cathedra from Jasper's seat on the then-pink couch (30). What were little more than the ordinary conflicts between parents and children over friends, dating and staying out were transformed by the political self-consciousness of the home into tests of theory (anarchism works or it doesn't), praxis (claims and counterclaims of fascist behavior), or both.

Perhaps for Jasper the incompatibility between this parental authoritarianism and theoretical anarchism was less apparent than it was for his kids (it was, after all, his technical definition of "true" in "surrealism is the anarchy of true love" that reconciled for him surrealism and his inescapable puritanism); but a more concrete analysis of the situation would have obviated the entire issue. The existence of an anarchist household in Cleveland Heights in the 1960s was an inconceivable delirium (the very *family* whose headship Jasper claimed—endlessly and in full voice—was a legal creature of precisely that *state* anarchism most resolutely opposed). To have achieved it, Nancy and Jasper would have to had freed themselves from the reproductive mechanisms they internalized in learning the rules in their parents' homes (making the rules theirs, laying them on their friends, relations [as in turn, Denis on Christopher, Peter {and in his turn, Randall on Chandler, Kelly}]), mechanisms insuring the reproduction in their own children of none other than their fraction of that class-structured society dependent on the capitalist exploitation of labor supported—*above all else* (as the fate of labor in our time reminds us)—by the state. Only one look at the fidelity of their reproduction of their parents' living rooms is required to understand the limits of their (doubtless heroic) attempts and their inevitable failure: a family is not a universe but an open system indissolvably whose with the society that simultaneously sustains and is sustained by it. When that society is distorted by the private ownership of the means of production for the sake of profit, institutions of social reproduction (for example, the family) will be similarly distorted. Since the state is "the chief instrument for permitting the few to monopolize the land, and the capitalists to appropriate for themselves a quite disproportionate share of the yearly accumulated surplus of production," no Wood doubts that it has to go; but it had not gone when Denis and his brothers were growing up; and the attempt to act (in any way) as though it had only further distorted a living under adequate stress from the torsions of the capitalist state.

No one understood this more clearly than Errico Malatesta, whose slim *Anarchy*—originally published in 1891—remains, at least for Denis, the most eloquent exposition of anarchist objectives. We quote, for example, apropos the foregoing, these paragraphs from the concluding pages of *Anarchy:*

> That's all very well, some say, and anarchy may be a perfect form of human society, but we don't want to take a leap in the dark. Tell us therefore in detail how your society will be organized. And there follows a whole series of questions which are very interesting if we were involved in studying the problems that will impose themselves on the liberated society, but which are useless, or absurd, even ridiculous, if we are expected to provide definitive solutions. What methods will be used to teach children? How will production be organized? Will there still be large cities, or will the population be evenly distributed over the whole surface of the earth? And supposing all the inhabitants of Siberia should want to spend the winter in Nice? And if everyone were to want to eat partridge and drink wine from the Chianti district? And who will do a miner's job or be a seaman? And who will empty the privies? And will sick people be treated at home or in hospital? And who will establish the railway timetable? And what will be done if an engine-driver has a stomach-ache while the train is moving? . . . And so on to the point of assuming that we have all the knowledge and experience of the unknown future, and that in the name of anarchy, we should prescribe for future generations at what time they must go to bed, and on what days they must pare their corns.
>
> If indeed our readers expect a reply from us to these questions, or at least to those which are really serious and important, which is more than our personal opinion at this particular moment, it means that we have failed in our attempt to explain to them what anarchism is about.
>
> We are no more prophets than anyone else; and if we claimed to be able to give an official solution to all the problems that will arise in the course of the daily life of a future society, then what we meant by the abolition of government would be

curious to say the least. For we would be declaring ourselves the government and would be prescribing, as do the religious legislators, a universal code for present and future generations. It is just as well that not having the stake or prisons with which to impose our bible, mankind would be free to laugh at us and at our pretensions with impunity!

We are very concerned with all the problems of social life, both in the interest of science, and because we reckon to see anarchy realized and to take part as best we can in the organization of the new society. Therefore we do have our solutions which, depending on the circumstances, appear to us either definitive or transitory—and but for space considerations we would say something on this here. But the fact that because today, with the evidence we have, we think in a certain way on a given problem does not mean that this is how it must be dealt with in the future. Who can foresee the activities which will grow when mankind is freed from poverty and oppression, when there will no longer be either slaves or masters, and when the struggle between peoples, and the hatred and bitterness that are engendered as a result, will no longer be an essential part of existence?

This sort of talk (sane and gentle though it is) is strongly resisted by states of whatever persuasion (as indeed, since it threatens their existence, they are well advised). Malatesta spent nearly half his life in exile, much of it in England. In Peter's painting—taken from one of the few photographs of Malatesta readily available—Malatesta stands outside the Bow Street Police Court in London (Bow Street prison Peter imagined [that's supposed to be barbed wire along the top of the wall]), where he appeared in 1912 on a criminal libel charge. His conviction drew a three-month prison term (Peter imagines that Malatesta is being released) and a recommendation for deportation (which, thanks to mass demonstrations in Trafalgar Square, did not take place). The discrepancies in the description of what is a minor moment in a tumultuous and significant life are less interesting than the dramatization within it of the inevitable and ultimately everyday quality of the confrontation of men of free will with the police power of the state (the *Malatesta* is thus history painting in the revolutionary manner of David).

Peter has called *Malatesta* a religious painting ("The anarchist movement is religious—isn't it?" he has wondered) in contradistinction to most of his others (such as his *Genovevo de la O* [64]), which he regards frankly as works of propaganda in which, aside from any artistic value, his essential interest is in getting people to ask about their subjects and so be led to "read and think about changing the world."

Evidently the *Malatesta* works on this level as well.

The Round Lamp

Cool is the light of this lamp, like that of the chrome of neck and base, like that of a streetlight, like that of the moon, which—when the lamp alone is on in the room, and the neck is dematerialized by the brightness of the light—it often seems to be.

This effect is an attribute of the "shade," a fourteen-inch diameter sphere of, originally, frosted glass (from Yugoslavia), but now a translucent plastic (from Mexico). Truncated so near the base that a quick glance fails to detect it, this simply *sits* on what is, in effect, a large harp wing flattened into shell. The power cord runs down inside the neck—a chrome-plated steel tube—and out through the base—a chrome-plated steel skin (shaped like the top of an onion) encasing three heavy cast-iron weights.

When it's plugged into that West Virginia coal and switched on—*moonrise!*

rules
176. "Don't play with it—you could knock it over" (Chandler, kAPP, tPRO).
177. "Don't touch it unless to turn it on or off" (Randall, ktCON).
178. "Don't take off the lamp shade and use it as a space helmet" (Chandler, kAPP, tPRO).
179. "Don't move it" (Randall, tCON, tPRO).
180. "Don't breathe on base" (Randall, tAPP).

Gem, precious metal, poor man's stainless steel: the lamp by the couch (34) . . . *passes* for modern. Dispensing with the form of its traditional shade, replacing its linen with plastic, this above Mies van der Rohe's lounge chair *is* modern. It even has a little of that "self-conscious theatricality"—balancing a sphere on a finger!—of Achilles and Piergiacomo Castiglioni's Arco lamp, coveted even in the act of purchasing these more modest standards (34, 42). Otherwise: no difference. Same anamnesis of an economy (evidently!), same classicizing elegance, same revolutionary romanticism (exposed metal), even the same sort of story about the kids knocking it over (only this was before Chandler's time), Randall scuttling around on the floor, Denis watching him, and before he knew he was gone *C!RA*S&H!!* Randall among the shards Denis petrified with terror . . .

There is a world at floor level for children that *recedes*—literally but also metaphorically—*as they stand up*. By the time they are really using the light: "I don't remember that rule. How dumb! Who would put their mouth down there anyway??"

XXVIII · THE CONTROL CODE

Despite Randall's *don't move it* (RULE 179), the round lamp *is* moved. Certainly it is moved at Christmas when the tree reaches into the space it usually occupies (then it is moved over), but on occasion it is also moved . . . into the dining room. This is admittedly not often (it is whenever candles would be inappropriate, but the thought of using the overhead light is too repulsive) but it is often enough. Where is the control rule that specifies this behavior?

As we know, it is not here. Not only does it not appear among the rules for the lamp, but rules like it do not appear among the rules in general. Furthermore, among the codes under which we have grouped the rules, that of control is the most impoverished; and its rules are often weakly embedded in it, gravitating more strongly toward those of appearance or protection. Such differences follow from the different ends aimed for by the different codes (as, outside the room, the aim of the highway code is to promote *physical* safety, and that governing sexual relations, a certain *moral* order). The protection code is ontic in its ends. More particularly, it is concerned with the *continuity* of things. The goal is preservation, and the rules constitute a preservative medium. They amount to a kind of brine, vinegar, formaldehyde: they are an embalming fluid. The room is pickled, the kids mummified (*don't touch, be careful, please don't move them, don't take off the covers, NO SHOES ON THE COUCH!*). The appearance code is phenomenic in its ends. It is concerned with how things look (in the kitchen and bathroom this is extended to how things smell). The goal is ostensive, and the rules constitute an ostensive medium. They amount to a kind of façade, costume, cosmetic: they are packaging. The pickle is in a pickle jar, the glass sparkles, the label is neat; the mummies are smoothly bandaged, the cartonage is firm, the encaustic portraits free of brush marks (*don't breathe on it, don't put your hands on the wall, don't leave anything on the cabinets, keep fingers off, don't get the couch dirty*). The control code is telic in its disposition. It is concerned with what things do. The goal is perseverance, and the rules constitute instructions for maintaining the room in a persisting state. They amount to a kind of recipe, handbook, program: they are an operating manual. The pickles will be not only edible, but nutritive; the kids' bodies will not only resist decay, their personalities will be perpetuated (*close the door if the furnace is on, when you close the door turn the handle, stairs are for going up and down, put records back where you got them, treat it like a chair, not like a toy*).

From the perspective of the control code, the roles played by the other two are silly (use necessarily results in wear [and *what is the point of buying a couch unless you're going to use it*]) and superficial (*if you worry about marking the risers, it takes all day just to get from one floor to the other*). This perspective, moreover, conforms to our myths about the *deep* (it's what's inside that counts) and the *essential* (use function). The mystery thus becomes (1) why so few of the rules can be coded CON—only marginally more than a third (whereas better than half receive an APP), (2) why so many of those that can are powerfully implicated in other codes (as *treat it like a chair, not a toy* is more concerned with the appearance of the kid than the use of the chair), and (3) why so few *specify* ends or means (most no more than interdict *forbidden* end or means, by, in fact, a ratio of five or six to one, and this with the most inclusive definition of positive specification [thus we accepted *keep your friends from dingdonging them too much* simply because it wasn't put in the negative]). We have insisted elsewhere that this respects an economy (*it is easier to say what a couch is not than what it is*) that in turn contributes to a powerful freedom (*everything not forbidden is implicitly permitted*), but this in turn is subject to an economy, that of imagination. In comic strips this is usually one of forbidden actions which have not been explicitly interdicted (such as that of Bill Watterson's Calvin who, interrupted pounding nails into the living room coffee table by his mother's "Calvin! What are you *doing* to the coffee table?" responds, "Is this a trick question, or what?"); but while this sort of imagination doubtless can result in that positive modeling of behavior we have associated with highly authoritarian households ("a coffee table is to put a coffee cup on—period"), it is evidently not the primary territory claimed by the child's imagination (whatever it may be for his parents). Indeed, imagination would seem to be ordinarily in the service of internalizing and building on precisely the behaviors *enacted in the home* (as when Calvin's friend Susie models playing "house" with "Okay . . . First, you come home from work. Then I come home from work. We'll gripe about our jobs, and then we'll argue over whose turn it is to microwave dinner"). The example is pertinent. Although apparently far removed from the simpler problem of how to use a couch, in fact the couch is a prop in the scene Susie describes (thus, a more detailed description might begin, "Okay . . . First you come home from work and collapse on the couch") *and* a product of the very labor that prompts it (that is, the work from which Calvin and Susie are to come home is justified, in part, by its ability to purchase and maintain the couch). Thus the positive behaviors that we might term "couch habits" will develop as a convergence of imaginative imitation of existing behavior (observation, mimicry, role-playing) *and* the discovery and invention of novel behaviors constrained explicitly by the don'ts of the rules. It is the imaginative imitation of observed behavior that *precludes* (interestingly enough) the pounding of nails into the coffee table by *in*cluding hammer and couch use in distinctive "hammer habits" and "couch habits"—in the contexts of which novel possibilities *will be* unfolded (but usually within the "habits" [*Calvin and Hobbes* unfolds in the space left open by the "usually"]). It is this simultaneously inclusive and evolutionary aspect of behaviors that renders them elusive subjects for rule formulation. The veritable *use* of the object, caught up in the innumerable agendas of those living the room, eludes every net cast over it. To any catalogue is inevitably appended another role played by the object at another time, or under this or that special condition, perhaps initially ad hoc, but now routine—"Oh, I forgot about that."

Exactly. The use so unimaginable (the glass top of the Mies van der Rohe table *on* the couch) becomes self-evident (but only when the table must be gotten off the floor for waxing). In the throes of the evanescent agenda . . . every meaning is altered. And yet, the *meaning* of the couch cannot be just one or another of its meanings: it is all of them, unedited by the heavy hand of "ordinarily" or "most of the time." Not only do these qualifiers themselves slip and slide over time, but as we know are *at any time* radically relative: thus, *don't leave anything on the cabinets* (RULE 101), except, it turns out . . . the daily mail.

Oh. I forgot!

But it is not just a matter of forgetting: *the plant sits there all the time,* and it soon is evident that no rule could be remembered capable of covering even the most recurrent exceptions. *Yet the rule must be memorable.* No matter the nausea implicated in the reiterated negativism of the don'ts: *don't leave anything on the cabinets* is significantly easier to remember than *don't leave anything on the cabinets, except the plant stays there all the time, the mail goes there along with things that need to be taken to neighbors (like Ann's book) or given to Martie next time she comes, as well as at Christmas the six tongue depressors that Chandler decorated for Denis a long time ago and now and then Denis's coffee cup and some other things.* But if the memorable rule is constantly violated, what's the point?

Just this: to direct, not dictate behavior.

Life is too contingent to reduce to aphorisms. Yet among the contingencies paths *can* be indicated. Home rules rarely attempt more, especially those devoted to maintaining it in a persisting state. It is for precisely this reason that so many rules annul, modify, or second guess other rules—because none is competent alone in the contingent universe of the living. Hence: *close the door if the furnace is on* (RULE 21) but *open the door for fresh air* (RULE 23); hence: *don't play with the bells* (RULE 36) but *keep your friends from dingdonging them too much* (RULE 38); hence: *don't leave anything on [the newel post]* (RULE 60) but *take up things on the newel post when you're going up* (RULE 59); hence: *put records back where you got them* (RULE 112) but *unless you don't know where they go* (RULE 114); hence: *please don't move them* (RULE 165) but *we move them at Christmas* (RULE 166). Even more illuminating are the rules offering up internal evidence of their failure (*don't wind it up and down like we sometimes do* [RULE 88]) or their failure compounded by further nullifications (*don't play the records too loud: if you do make sure Bill's window is shut* [RULE 171] but at the same time *unless Den's working, then we can't play it too loud* [RULE 172]).

"At least be consistent," parents are enjoined, but the demand for the memorable leads to simplification; and in the contingent world of the living, simplification leads to inconsistency. The result is inevitable: a congeries of ad hocist pronunciamentos hardened by time and obduracy into inconsistent simplisms.

Miraculous consequence? The room is the entelechy dreamed of by the control code. Everything works. The room not only persists, it reproduces itself.

The Plant by the Speakers

Every week a demonstration of the monsoon: after Ingrid waters it, the turgid stems of the peace lily stand erect, waving their leaves in the air like flags; but as these emit water (and it is evaporated from the potting soil), the plant loses turgidity, it wilts, it droops, and its lower leaves rake their outer faces on the floor like dust rags. If it's really hot and fans are roaring day and night, or cold and the radiators all but boiling, then the plant collapses on itself like a dishcloth huddled in the sink.

It's another peace lily, the mother, in fact, of that on the floor in front of the easel (18). Drooping, it's not eighteen inches tall, twenty-eight from side to side; but full of itself, it reaches six or seven inches higher, stretches itself another four or five from side to side. The pot is terracotta, nine across and seven inches deep, inside a plastic saucer made to look like pottery. This is one of the Clay Beaters from the A. L. Randall Company of Chicago, Illinois.

But with the regularity of the monsoon, the Ingridian rains return. Then the stems leap upon the air!

rules   181. "Don't hurt it" (Denis, kAPP, tPRO).
        182. "Don't abuse it" (Randall, kAPP, tPRO).
        183. "Don't be mean to it" (Chandler, kAPP, tPRO).
        184. "Don't step on it" (Chandler, Randall, kAPP, tPRO).

It's a Justinian digest: all the rules for the plants are compiled here. Denis reprises his *don't hurt the plants* (RULE 84), and Randall his *just like all the other plants—don't abuse it* (RULE 99). Chandler's version of Randall's *all plants, don't kill or injure—be nice* (RULE 86) is his *don't step on it* and *don't be mean to it*. Nature, alive in the room! respected!!

Who asked?

43

Exactly! The minidrama of the monsoon spells out the limits of this brotherhood of Man and Nature. Daily we can mark the dependence of this Nature Tamed by the increasing degree of wilt. When it becomes intolerable (when it offends our sense of decorum) we—generously (little gods) . . . give it a drink.

Noblesse oblige!

The Koala Bear in the Plant

It's not visible from even a few feet away, this little tan koala clinging to a branch of brown. It's a symphony of plastics, the bear's body being molded from one of extremely high density, its "fur" of a plastic dust, eyes, nose, and branch of a high-grade polystyrene. The bottom two inches of the branch—free of the marks of bark—descend to a sharp point meant to be shoved into the earth. A trinket of scant worth, on inspection it nonetheless displays the richness of a sophisticated petrochemical technology, as on inspection the parted leaves of the peace lily (43) display a sluggish Australian arboreal marsupial shrunk to the size of a thumb.

Its eyes shine, whatever.

rule   185. "I think—don't take it out and put it someplace else" (Chandler, tcon).

When asked, "Why not?" Randall shrugs his shoulders saying, "I don't really know." The tentativeness of Chandler's rule, the diffidence of Randall's response, unmask the marginality of the object in the ecology of the room. It is hardly dizzying, evidently unnecessary, far from essential. And yet . . . and yet . . .

Until her death, Ingrid's Aunt Irmi—alone in Bremen—sent the Woods incredible packages at Christmas, not large, but stuffed with *things*: walnut shells gilded and strung for the tree, brightly tinted marzipan fruit, boxes of *pfeffernuse, lebkuchen,* German chocolates, ornaments, and small toys—each in its own wonderful paper—and the koala was one of these, slipped the Christmas of its arrival into a pot and preserved there, token, remembrance, memorial, relic (the room is a shrine, the room is a graveyard).

Denis and the kids met Irmi only once. She wasn't a sluggish Australian marsupial, but her eyes shone, whatever.

Standing in the dining room at the knickknack shelf just inside the archway to the living room, Denis and an acquaintance were admiring the badge Denis's great-grandfather had worn as a postman in Boston.

"So," the acquaintance said, "some of these things have family associations?"

Where to start? With the white walls and the floors without carpets? With the bells on the door? The couch? With the Yunkerses on the walls, the Strand? With Peter's paintings? With the taste for folk art? With the interest in Mexico? The wooden car on the hearth? With the commitment to modernism? To the clean? With the division of the floor into living room, dining room, kitchen. With the—it was *all*, really, *all or nothing*.

"Some more than others," Denis acknowledged, steering the conversation into less . . . *roiled* waters.

The Front Radiator

It is easy to overlook the radiator, easy to let it be swamped by the many things clamoring for attention. It is easy to lose it in the space below the window, to let the wall absorb it. Unless it's winter and the snow is falling (or better yet—it's sleet) when the radiator becomes the very raison d'etre for being in the room, for turning the rocker around to face the window: to watch the weather across the toasting toes.

Otherwise this is another forced-circulation hot-water radiator like the one (9) beside the stair, longer—five and a half to the other's three and a half feet—but otherwise hard to distinguish, betraying the same history of painting and decay, the same rust and metallic gray . . .

rules   186. "Don't stand on the radiator" (Randall, ktPRO).

187. "Don't put stuff on it and leave it there" (Randall, kAPP, ktcON).

In fact, "put them on a radiator," is a winter's common cry, as kids and Den (and sometimes Ingrid) strip from a morning, afternoon, or night's hard sledding out behind the Project or up Dix Hill and yet another looms—imminent—without a change of clothes. "Put them on a radiator where they will dry," and there is no thought of staying with them—you leave them there, along with the rule.

How to construe this?

The rule says, "Don't put your Lego spaceship, book, Swiss army knife, cup, pencil,

schoolbook on the radiator and casually leave it there." It says, "Don't be a jerk," while saying what a jerk is (that is, one who would casually leave stuff on the radiator). The rule's role (again) is to direct, not to dictate behavior; and any purpose understood as reasonable is capable of vacating it (that is, one underwritten by thought [it is thoughtlessness that is jerky]):

"I'm just leaving it there while I get a drink."

"Okay. But don't forget it."

At the same time the wet clothes and the wet weather together turn one of Irving Goffman's keys (yes, we have transposed from his musical metaphor to one of control [wonderful: the one in the other {incidentally: a surrealist game}]). That is, as happens at Christmas, the double wetness induces a phase-shift whereby, for the duration, everything is altered. Whereby the embodiment in the room of another refusal: *no clothes drier* (as this is written Ingrid is at a neighbor's . . . drying clothes).

That *don't stand on the radiator* is coded ktPRO while *we can't stand on it* (RULE 53) is coded simply kPRO acknowledges that in falling *here* the kid would more than hurt himself. Flailing and windmilling he could smash, knock over, or step through the wicker rocker (35), the couch lamp (34), the coffee table (65) . . . who knows what all.

The Side Radiator

Every year Denis decorates the wrappings on his Christmas presents for Ingrid. Now and then these have told stories. One year these were the adventures of a little horse who managed to wriggle off the Christmas tree and had to spend the rest of the year evading mops and brooms, feet and vacuum cleaners. One picture shows the radiator from the perspective of this little horse about to take refuge beneath it. The radiator is massively present as the piece of plumbing it is, identical to those at the foot of the stair (9) and along the front of the room (45) in everything but length.

Wherefore for the little horse: terror of the radiator brush.

hand-off   3. "Same as others" (Chandler, Randall, to RULES 52, 53, 186, 187).

Here, three points. The first is the way nothing in the room is exempted from the signifying activities of those who live it. Moreover, as our example demonstrates, these need not depend on *prominent* affordances made by the object for the room. Thus, the narrative here exploits the radiator in neither of its classic roles (as heat, as surface); rather, as in mythic thought, the narrative draws correspondences between two domains—that of the objects in the room (ornament, radiator, vacuum cleaner) and that of human life in the world (heroes, refuges, threats), so that the semantic field to

which both belong is enriched, nuanced, more precisely defined: forced-circulation hot-water heater is revealed as the overhang of a cliff, the ceiling of a cave.

Not that this space was unknown to the room. In fact the radiator brush knew it well, or, more precisely, one of the two radiator brushes. This is the second point, the way nothing in the room is exempted from participating in the circulation of dirt, each object conjuring up its own attendants in the mirror world of the cleaning, here, beyond the long-handled brush for getting under and beneath the radiator—originally encountered cleaning the floor (8)—another, round brush, for getting in between the tiers or vanes of the hot-water tubes. Here we recognize the threat of the dusty, the desiccated term of that dirt (*mud/ grime/ dirt/ soot/ dust*) that is the inevitable end-state of all cultural degradations (for the room: no difference between ghost town and jungle ruin).

Third point: the way everything is subject to locative qualification. Although the three radiators (8, 45, 46) differ as steps, stools, and tables in length alone, *take items on the radiator upstairs when going up* (RULE 54) applies uniquely to the one at the foot of the stairs, where its location makes of it a staging area for the ascent to the second floor (the room *is* a hodological space). It is similarly the location below the windows on the street of the radiator along the front of the room that makes of it a hassock when the winds blow free. And *its* location at the ends of the earth that makes of the radiator between the speakers a likely spot for a game horse to hide.

XXX · CHRISTMAS

Incandescent occasion: the rules go up in smoke.

A freshly cut evergreen tree reaching to the ceiling is installed in the corner otherwise occupied by the left speaker (38), the Lacandon drum (39), the round lamp (42), and the plant by the speakers (43). To accomplish this the room is substantially altered. To bring in the tree, the storm door (1) is released from its spring-and-chain door stop and opened flat against the outside wall and the door (2) opened as far as it will go. To make a passage for the tree, the rug ordinarily between couch and stair—the Navajo or the kelim—has been rolled up (8), the couch (30), and table (65)—cleared of everything (66, 67, 68)—have been moved against the small record cabinets (23); and the David Rowland chair (29) shoved into the dining room, along with the round lamp (42) and the lounge chair (53). The left speaker has been turned to face the corner. Increasingly it plays Christmas records (27).

For the next couple of days the room is a shambles. Glass balls and strings of lights clutter the couch. Strings of popcorn, cranberries, and paper rings stretch into the dining room. The coffee table is mosaicked with straw crèche figures, wooden animals, tin soldiers, lace snowflakes. The floor is littered with packaging, boxes. At the same time the mantle (62) is cleared (63), and one of the throw pillows (36), the plaited fan (59), the wooden car (60), Denis's collage (32), the Cherokee basket (24), the Lacandon drum (39), and the rolled-up rug (8) are carted upstairs. The throw pillow Martie made for Christmas is brought down and so is the Christmas cover Ingrid made for the other pillow (36). The rug is replaced by the braided one usually in the dining room, Denis's collage by one of the nativity he and Gary Richmond made in the fourth grade, the wooden car by a papier-mâché reindeer Martie gave the Woods, and the plaited fan by five Christmas stockings, two of them Denis's, a huge one Ingrid knitted him and a smaller one from his earliest remembered Christmas. A Mexican pottery crèche aggrandizes the mantle. Santas, trees, snowmen sprout on the meeting rails of the inner sashes of the front and side windows (47, 50). On the cornice of the porch window (17) wooden angels and evergreens appear, and later, sprigs of actual evergreens. Ingrid tucks more of these into the crown of the king (55), along the back of the mantle, on the box tray of the easel (19). She runs great swags of it along the banister (11). Before long the round lamp returns to occupy a place four feet east of its usual one. The coffee table slips closer to the center of the room, pushing the couch further toward the stair. Couch lamp (34) and rocker (35) also migrate. A tin crèche mounted on an old potato chip display rack materializes on the coffee table, where now and then it is joined by a candelabrum. During the first weeks of December wrapped presents pile up beneath the tree. Christmas Eve a plate of cookies and a snifter of brandy are left on the coffee table. Other presents, wrapped and unwrapped, materialize during the night.

This great raveling begins to unravel the next morning. Elaborately wrapped presents are ripped open, unexpected largess displayed and admired, toys assembled, traditional foods eaten *in the living room* where the Woods sprawl for hours among a luxurious debris of gifts and wrappings, beginning to pick up only with the onslaught of the company coming for dinner, and then only partially, books, toys, records, foodstuffs in defiance of all rules not only permitted but encouraged to stay on the floor, under the tree without special license, but spreading around it, too (even in the circulation space between the stair and the couch), in a lavish spectacle of (tasteful) prosperity that infects every surface including the ceiling (69), where the light from the star on the top of the tree flares when the other lights are low. Gradually—this can take weeks—the toys slip upstairs, the food is eaten, books and records read and unread, listened to and not listened to are shelved or returned or passed on, the basket and collage and drum come back, the tree comes down, is chopped up for mulch, the decorations once again are shoved to the back of the closet in the middle room: the living room is restored. On that day Denis always airs out the room (you can't open the window with snowmen and Santas on the railings of the sashes) and plays Miles Davis's *Blue Haze.*

The room swings, once again, to the rules.

Several points. First of all, everything's different, nothing's changed. With trivial exceptions, the paradigms and themes sounded in the room the rest of the year are sounded at Christmas. Has the paradigm *folk/ popular/ elite* been saturated? Then it will be saturated on the tree (*Bolivian straw crèche figures/ fat blue, green, yellow, red light bulbs/ hand-blown art-glass balls*). How about the paradigm *machine/ man/ nature?* Even without invoking the tree itself we find *the ornament hooks/ hand-crocheted snowflakes/ sea-shells.* More iffy (and this time of the home): *wood/ low-fired clay/ high-fired clay/ copper alloy/ brass/ tinned steel/ stainless steel/ silver* (this is the economy of substances invoked by the bells). Nonetheless, missing only the stainless steel term, and without leaving the tree: *little wooden train/ terra cotta plaque/ porcelain angel/ Marie's three copper kings/ the brass angel/ the Mexican tin crèche figures/ ————/ the Towle silver plaque.* Or take *crude/ fine.* On the tree we find an angel Randall made from a clothespin, but also a tiny articulated pewter Santa Claus. Is the living room identified with Denis's side of the family? Then so is the form of the celebration, presents being unwrapped in the living room Christmas morning around a tree lighted with bulbs (whereas the Hansens generally unwrapped their presents in the dining room Christmas Eve around a tree lighted with candles).

If nothing is changed, how do we know it's Christmas? Because everything is phase-shifted. Though Nature is *not* admitted to the room with the tree—which not only is dead but in the end little more than a surface for the display of a simulacrum of Culture—the assertion of the Natural achieved by the plant displaced (43) *has been enhanced* any number of degrees:

"Oh, what a pretty tree!"

"Don't you just love the smell of the real ones?"

"It has such a perfect shape!"

This not only causes a shift in the fundamental paradigm structuring the room—that is, *Nature/ Culture* (most manifest in the sudden tolerance for pine needles)—but literally shifts the furniture (necessarily, since furniture and paradigms are merely different ways of saying the same reality). This induces ripples throughout the room in every aspect of its being. A few of these were caught in the net we set for the rules: *but we move them at Christmas* (RULE 166), for example, no more than acknowledges the threat of the tree's branches to the face of the left speaker (38); and *Denis puts the (chooses the) easel items except at Christmas when Ingrid decorates* (RULE 89) may also be taken at face value. But *cover speaker if you're doing work around it that might allow stuff to fall on it* (RULE 98) is a tribute to precisely this decorating, specifically of the banister (11) with swags of greenery. What exposes the phase-shift are all the qualifiers: *but* we move them, *except* at Christmas, *if* you're doing work . . .

We have implied that phase-shifts like this are brought about by something like the keying of Irving Goffman. Here is the full definition of keying from Goffman's *Frame Analysis:*

> a. A systematic transformation is involved across materials already meaningful in accordance with a schema of interpretation, and without which the keying would be meaningless.
>
> b. Participants in the activity are meant to know and to openly acknowledge that a systematic alteration is involved, one that will radically reconstitute what it is for them that is going on.
>
> c. Cues will be available for establishing when the transformation is to begin and when it is to end, namely, brackets in time, within which and to which the transformation is to be restricted. Similarly, spatial brackets will commonly indicate everywhere within which and nowhere outside of which the keying applies on that occasion.
>
> d. Keying is not restricted to events perceived within any particular class of perspectives. Just as it is possible to play at quite instrumentally oriented activities, such as carpentry, so it is also possible to play at rituals such as marriage ceremonies, or even, in the snow, to play at being a falling tree, although admittedly events perceived within a natural schema seem less susceptible to keying than do those perceived within a social one.
>
> e. For participants, playing, say, at fighting and playing around at checkers feels to be much the same sort of thing—radically more so than when these two activities are performed in earnest, that is, seriously. Thus, the systematic transformation that a particular keying introduces may alter only slightly the activity thus transformed, but it utterly changes what it is a participant would say was going on. In this case, fighting and checker playing would appear to be going on, but really, all along, the participants might say, the only thing really going on is play. A keying, then, when there is one, performs a crucial role in determining what it is we think is really going on.

Evidently the match is imperfect, but each criterion is in fact satisfied. Certainly the transformation of the room is systematic, and across materials already meaningful. Certainly everyone acknowledges this alteration which radically reconstitutes their behavior (the kids are *incapable* of coming downstairs [15] Christmas morning until Den and Ingrid have). And certainly cues are abundantly present to mark the temporal and spatial boundaries of the transformation. However, what Goffman's fourth criterion implies for our very different domain is less certain—though as we shall see, keying is not restricted to Christmas, or even holidays—and it is even less certain how his fifth applies, except insofar as it is no more than a higher threshold than his second for asserting the presence of a key change. Nonetheless, his definition allows us to see Christmas as one of many similar events and not as, for example, a unique intrusion of the sacred.

Thus, as we have already seen, the double wetness of wet clothes *and* wet weather can force the room to modulate from the key of dry to that of wet (thereby temporarily vacating the rule keeping things off the radiator [RULE 187]); and if this seems transcendently trivial next to the momentous change brought about by Christmas, it nevertheless meets the conditions set for a change of key. In general, motivations for changing keys will fall into five groups. Doubtless most significant are the *celebrations*. These are marked by the regularity of their recurrence, the resources they mobilize, the extent of their redefinition of the room. Christmas casts its aura over everything, but the salience of Easter, Valentine's Day, and the many birthdays should not be underestimated. The transposition was particularly acute when the kids, much younger, had birthday "parties." These were denoted by the throngs of guests, the streamers and balloons, the fits of anger and hysterical play, the (leftover) cake and ice cream. Marked by their unpredictability and comparative uniqueness are *parties*. Denis and Ingrid have held only one "party" in this house, but have had couples over for dancing twice. Because the participants are adult (that is, civilized [that is, not barbarian]), the issue of rules is not raised, but the room is more radically rearranged—if more transiently—than by Christmas (the role of the floor is enhanced [you can't dance on furniture]). The transpositions attendant on *illness* are the most subtle: the room slips on slippers and a robe, the couch declines to no more than a bed (the room drowses [it is as if a curtain were pulled, even in sunlight]). *Weather* too can modulate the room, from the key of dry to that of wet, from that of hot to the key of cool. This last is overheard in the rules where a phase-shifting qualifier pipes up in *close the door if the furnace is on* (RULE 21), but it is most clearly heard in the composition of the room we inventoried: screen (1) and fan (61) in the absence of rugs (8) mark a room in the key of hot. The key of cool is noted in the storm door inaugurating our description of the bringing in the tree, as well as in the presence of a rug between couch and stair. Dramatic shifts to very distant keys is the mark of the *cleaning*: intrusive ladders in the living room (as when Ingrid washes the windows); windows open . . . when the heat is on (as when airing out after Christmas); furniture shifted . . . but not by celebrations or parties (as when Ingrid waxes the floors). The point? To maintain the room . . . in the key of clean.

But if this is the point, why ever leave it? Why use the room, why get it dirty (why not use it like the parlor)? Why unleash a birthday party in it? Why move all the furniture around, put up decorations only to take them down, endure the needles of the tree . . . in everything? We know the answer to this question. It is not that the room is *used* for these things, not that they take place *in* the room, but that they *are* the room, the living that the room *is* . . . in another key.

And what key is the room in? Perpetually modulating from one another, it is hard to say. It is not tonal, but the room takes no care to be atonal either. Not reducible to a given sequence (history, the maturation of the kids assures us of this), it most nearly approximates the serialism of Anton Webern, whose idea of multiple fields of distribution—released from the implications of one original *Grundform* (of one reified crystalline Room)—alone seems capable of modeling the continuously interacting structure and freedom, invention and tradition, that are the room, that are *Christmas* even, with its accumulated burden of two millennia, nevertheless . . . ever fresh.

The Front Windows

At night the street light strides through the windows at the door, but it skips through the glass of the front windows, slapping the south sides of the muntins in the outer sashes, teasing them to life from a local shadow deeper than the shades of night. There is this ebony edge, a fillet of ivory, and through the glass a dappling of charcoal and ash, then the ebony, the ivory, the ash . . .

Depending on how you count, this happens sixteen times, for twelve demiquarrels, six quarrels, and eight long skinny pentagons fill the frames of the outer sashes of the two double-hung windows, preempting with a play of light thirty-six square feet of white-painted papered plaster. Again, it's a contrivance of many parts: it is hard to be sure, but there are forty muntins in the outer sashes alone, and more than a hundred and fifty other jambs, stiles, rails, strips, stops, pulleys, cords, panels, and pieces of molding in what is, in the end, a fairly fragile machine. The inner sashes are glazed with single sheets of window glass thirty-two inches high and twenty wide. A simple molding smooths the transition from the stiles and upper rails to a wide mullion between the windows, and to jambs and header trim. A more elaborate molding reaches five inches below the apron, while over all looms a molded cornice above a six-inch frieze. Handles are inset in the lower rails of the inner sashes, locks on the meeting rails.

rules 188. "Keep your fingers off the glass" (Denis, ktᴀᴘᴘ).

189. "Make sure glass and screen don't get broken" (Ingrid, ktᴘʀᴏ).

190. "Don't open like *boom boom:* open and close continuously" (Randall, ktᴄᴏɴ).

191. "Don't try to talk through the window" (Denis, kᴀᴘᴘ, kᴄᴏɴ).

We know the windows in the room (4, 5, 17, 47, 50) do not bear the mark of the gem in equal degree, but as a function of the extent to which they will work. This is confirmed in the rules. Fourteen of the fifteen rules stripped from the inoperable windows (4, 5, 17) devote their energies to issues of appearance. Only two deal with protection, one with control (*don't spit on them* [ʀᴜʟᴇ 35]). In contrast, four of the six rules emitted by windows with operable sash (47, 50) concern themselves with matters of control, only three with appearance, and two with protection. Doubtless this reflects the order in which the rules were collected (as implied by ʜᴀɴᴅ-ᴏꜰꜰ 4), but there has also to be an understanding that *keep your fingers off the glass* is bound to be perfunctory when hands are swarming on the windows in the attempt to be heard (the rule says, *don't try to talk through* [the glass, open it if you want to be heard]), to be not heard (the rule says, *when playing a record close the window* [so Bill won't be bothered by the noise] [ʀᴜʟᴇ 194]), and to stay dry (ʀᴜʟᴇ 195). This is especially true for windows offering no purchase but that granted by placing the ball of the thumb, heel, or side of the hand against the glass beneath the meeting rail of the inner sash (that is, *on the glass*) and pushing up; or by wrapping the fingers around the bottom rail (that is *with the tips all over the glass*) and pulling down.

Unavoidably: grease on glass.

The Front Screens

When the inner sash is raised at night the space between the mullion and the jambs is no more than canescent until the moths attracted by the light mime the presence of a window. Surefooted, they thrum and crawl until someone—irritated—snaps the screen. The moths back, dart, and hover, but soon enough begin again their confident movements up and down the glaucous void.

A smaller bug might even make it in. The screens are aluminum and old, and the corrosion is accompanied by tears. But they are still tight in their frames, and the splines have rarely to be prodded back into their channels. The frames too are aluminum, as are the tracks in which they are held by springs at top and bottom. A latch,

intended to be lowered over the lip of the lower track sill, is secured by screws to the lower rail of the frame.

rule   192. "Don't let friends press their heads against it" (Randall, kAPP, kCON, tPRO).

Open the door for fresh air says the rule (RULE 23) and we know the same applies to the front windows, that they are open . . . to air out the room.

What can this mean? Only that the air of the home—that is, the air of Culture—is stale; that is, old, tired, exhausted; that is, too long exposed for sale; that is, polluted by the corruptions of civilization (etymology: urine). This is the Plutonian face of civilization: it spoils, it loses its elasticity, its spring. It requires periodically to be . . . refreshed, demands . . . recreation, needs to get . . . *outdoors* (resilient myth of Demeter and Persephone). Why not . . . *let it in?*

As with pets and houseplants (18, 22, 43, 58, 67), however, this outdoors will be a sham, the Nature it admits will be domesticated. It will be tamed . . . by the screens.

There will be an illusion of Nature (birdsong, cricket-shrill). But inside: *no bugs.*

The Front Storm Windows

Scritch!

Screeeeech-bang!

It happens every time. As the heat moves to the front of the house, Ingrid raises the screen *scritch!* to lower the storm window *screeeech-bang!* The *bang!* comes when the sash catches at a stop that no longer lets it slip on by. If it's the kids closing the window, it's *screeeech-bang! screeeech-bang! screeeech-bang!* until they give up, leave it open an inch, and close the window proper:

SkEEch, slslsl-chunk!

The *chunk!* is the sound of a wooden sash hitting the sill. It's not a sound made by an aluminum sash sliding in an aluminum channel, especially if it's been recently lubricated. There are three channels in each window—one is for the screen (48)—in a frame cut to fit within the outside casing of the window (47). In the spring and fall, both glazed sashes are up, held in place by spring-loaded stops that slip into holes drilled at regular intervals up and down the inner walls of the channels. In the summer, the inner storm is lowered in the day against the heat, but raised at night for the air; in the winter, it's lowered for the season.

Except when Denis plays *Blue Haze.* Then it's raised to air the room.

rule   193. "Close the storms if you're washing the porch" (Ingrid, kCON).

What are the storms that they should be closed when the porch is washed?

They are shutters, that is, movable covers for windows closed against bad weather. If today this is invariably meterological, it has often been human (etymology: to shut in, lock up, fasten with a bolt).

Remarkable achievement: to make them of glass (this is their luxury). A new material, the glass attests to their modernity (so does the aluminum [they are the Barcelona table {65} of shutters]). Transparent, fragile, easily broken (as the rules make clear), the glass also attests to the security of the neighborhood (one could literally *break* in [they are the unlocked door {7} of shutters]).

They are a sign of the times, one Philip Johnson saw with the greatest prescience: contemporary bourgeoisie will live in . . . glass houses.

The Side Window

Afternoons it almost hurts to look out this window so bright is the bounce of light from the wall of the house next door. Sixteen feet to the north, when the wall was white, it was even brighter. Now it is a pale bayberry, but even so it is blinding to try and read the paper against this brilliance. Hence: the wicker rocker—its light filtered through tree and porch—not the couch for reading in the early afternoon.

Other times the light is lovely, especially in the morning when it too is fluttered by leaf and branch. There's enough of it. The lower light is the second largest in the room, but a tenth of a square foot smaller than that in the door (4). Taken together with the lights in the upper sash—for of course there is the usual reticulation by muntins into quarrels, demiquarrels, and long skinny pentagons—it is the largest single window in the room, with two square feet on even the screen door (1). It's the usual contraption, similar in all but proportion to the windows in the front (47). This one is broader than those, more relaxed. A Gothic feeling to the thinner windows there gives way here to something Georgian, classical.

The light, of course, in its sparkling abundance, is purely . . . Mediterranean.

rules   194. "When playing a record, close the window" (Chandler, ktAPP, ktCON).
                195. "Close the window when it starts to rain, because the front windows are protected by the porch" (Chandler, ktCON, tPRO).

hand-off   4. "Same things as other windows" (Chandler, Randall, to RULES 26, 27, 28, 29, 30, 31, 32, 33, 34, 35, 79, 80, 81, 82, 83, 188, 189, 190, 191).

This window rises above the side radiator (46), which we saw stood at the ends of the earth. But far from sharing in its isolation, this window is the object of three rules specific to it (RULES 171, 194, 195), as well as a hand-off to nineteen others. Moreover, two of these rules (RULES 171, 194) tie the window to a relished activity—listening to music—which knots the window, the two speakers (38, 54), and the records (27) into a complex with affiliations not only to the universe encapsulated in the records (and through the act of listening, to the couch [30] and so on), but through the neighbor whose peace is to be respected by *not having to hear the records* (and whose home is visible through the very window which will effect it) . . . to the neighborhood, the city, the very idea of citizenship (here the closing of a window to spare a neighbor the burden of another's taste). The remaining rule (RULE 195)—in a lovely piece of collaborationism—raddles the window through the rain with the weather, the climate, the atmosphere, the hydrologic cycle.

The window is not just something you see through, it is something you see *with* (the room *is* a mathesis).

The Side Screen

There is the sense that everything but the bugs and the leaves makes it through the window—what we like to think of as air—but that this isn't so is something recalled every time a screen is cleaned: clouds of vegetable fibers, hairs, synthetic filaments, insect limbs, leaf parts, ash, soot, skin flakes, pollen, and dander rising with each stroke of the brush. Not that most of these don't pass: they do. The mesh is too crude to keep much out (it was woven with insects in mind), but a hair can cut its sixteenth inch to a thirty-secondth, and the mosquito tibia that snags can cut it to a sixty-fourth, and in time a kind of a felt develops that makes the screen every day . . . more visible.

Fiberglass screens like this don't rust or corrode, just tear and sag, but this one's done neither. The splines are still tight in their channels, and the screen is still tight in its tracks. Unlike the aluminum in the front windows (48, 49), this is anodized: light, tight, and white.

hand-off   5. "Same as the others" (Chandler, Randall, to RULES 1, 2, 192).

Who are the insects the screens are trying to keep out of the room?
They are the barbarians.
This is literally and metaphorically true. It is literally true that insects are widely perceived in our culture as irreconcilable enemies whose triumph will necessarily

entail our demise. This is a myth strongly promoted by petrochemical companies eager to sell insecticides: "At some future date, this may be an insect-free world. Until then, screens are important and apt to be costly," is how *America's Handyman Book* puts it, dismissing in the interest of dispensing with window screens a class of the phylum Arthropoda which—with more than 700,000 species—is by far the largest taxonomic division of the animal kingdom.

Metaphorically they are a prominent class of the phylum Barbariana, which in its other classes embraces rabbits, rats, snakes, kids, mildew, foreigners, wire grass, dry rot, male cats, the lower classes, bats, slobs, jerks, and bluejays, all of which are equivalent from the perspective of the paranoia that is Culture. The threat of the insect is that of any barbarian, whose terror derives from the termination he implies to the enterprise of Culture. Since Culture is a machine for teaching rules that have been taught, the intervention of a single generation of barbarians can mean the destruction of all that has been accumulated for millennia: if you can't read Homer, you may as well wipe your ass with him (etymology: barbarians can't speak Greek).

Evidently this is an hysteria (monthly visits of the exterminator help Denis and Ingrid keep it under control) but not one without importance. Culture may be a machine for teaching rules that have been taught, but these are rules that maintain the Culture *in its difference* from those of the barbarians (adults are *not* kids, the upper classes are *not* the lower, Americans are *not* savages [that is, black Africans, red Indians, yellow Viet Cong]). In Laura Ingalls Wilder's *Little House on the Prairie,* Pa:

> hitched Pet and Patty to the wagon and he hauled a tubful of water from the creek, so that Ma could do the washing. "You could wash the clothes in the creek," he told her. "Indian women do."
>
> "If we wanted to live like Indians, you could make a hole in the roof to let the smoke out, and we'd have the fire on the floor inside the house," said Ma. "Indians do."

Culture is this difference transmitted from generation to generation, a difference maintained *in tension* (hence C. P. Cavafy's, "Now what's going to happen to us without the barbarians? Those people were a kind of solution") but *as a generalized insect* (hence the screens, the exterminator, the can of Raid in the kitchen cupboard ["The Complete Bug Killer"], the Cutter's in the bathroom ["repels mosquitos, chiggers, gnats, biting flies, ticks, fleas"]).

The room: dust, mud, dirt, and insect free.

99.99 percent pure.

The Side Storm Windows

The long rains of November drive against the house, beat beneath the eaves, dance on the storm windows of its northern side. Sometimes the water slithers down the window in a promiscuity of twining snakes; other times it slashes at the glass with the feet of cloggers.

Bigger and anodized, these are otherwise similar to the storm windows (49) along the front of the room. There are three channels—one is for the screen (51)—in a frame cut to fit within the outer casing of the window (50). In the spring and fall, both storm sashes are up, held in place by spring-loaded stops that slip into notches cut at regular intervals up and down the walls of the track. In the summer, the inner storm is lowered in the day against the heat, but raised at night for the air; in the winter, it's lowered for the season.

Here the exception that proves the rule: nothing in the room is exempted from the signifying activities of those who live it . . . except for the storm sashes on the north window. Is this possible? Hardly. Denis and Ingrid scrimped and saved to have these installed, and they embody for them values of care, frugality, the husbanding of scarce resources, the very values embodied in their operation as well (that obsessive dance of windows, screens, storm windows, doors, and fans). But from the perspective of the rules: nothing. How can this be?

Certainly their location at the ends of the earth is not irrelevant, though as we know the window they protect (50) is not too far away to be the object of three specific rules and a hand-off to nineteen others. Yet this objection has the objection that if it is from the weather that the storms protect the window, it is from the kids that the window protects the storms. It is for this reason that there is no point here to *keep your fingers off the storms:* there's no way to get the fingers to them. Here it is distinctly relevant that the one rule collected for the front storm windows (49) approached it from the porch, that is, from the outside . . . as weather (RULE 193). As room rules, therefore, there is no point here to any of the hysterical interdictions of touching, smudging, breathing, spitting (that is, to the appearance code); or, for that matter, of pounding, hitting, doing anything that might break the glass (that is, to the protection code). As for the control code, at nine and eleven: too young to work the window, too dumb to care.

What is involved in operating the storm windows?

Connoisseurship.

What is required first of all is an appreciation for their physical character. To reach the latches in the inner sash—that usually lowered—it is necessary to raise the screen and lock it up (none of this demanded by a window sash). Then the latches must be pulled in—both at the same time—and the sash *powered* down (otherwise, because it's heavy, it may fall, and though the spring-loaded latches may stop it, it may still break [another threat not present in a counter-weighted window sash]).

That is, a storm sash is not a window sash (a little discrimination is required).

Once the machine has been mastered, it is required to know when to use it. As we know, the storm windows participate in a thermal farce, an elaborate collusion of windows, walls, doors, fans, storm sashes, and furnaces organized to maintain the range of temperatures in which the living unfolds. This calls for an understanding of the climate (subtropical), its expression in the weather (now hot, now cool, but always variable), and the role in its mediation of these components (windows, fans), both individually and in functional ensembles. It also calls for an awareness of their cost, especially in electricity and oil, and of the fact that these are paid for in the coin of a labor, few of whose satisfactions are inherent (terrible memory: Denis's father complaining about the money escaping with the heat through the bedroom window Denis kept open in the winter). Much of the fan and window dance can be attributed to the absence of air conditioning. The necessity of justifying this refusal to Randall and Chandler (many of whose friends do not suffer this deprivation) leads to an explication of the source of the *electricity* in West Virginia coal fields, of the *oil* in the Middle East, of the claims of miners, Arabs, OPEC and the UMWA, profit and pollution, exploitation and waste.

Nor have we done. Each of the windows is subject as well to the demands of its location. The front storm windows must be closed to wash the porch (because washing the porch involves washing the wall); and when the kids get older (as at this writing they have), they will be expected to close the *storm* sash on the side window when it rains (to protect the sill).

What is this if not *expertness in a subject requiring discrimination?* If not *knowledgeability, especially in recondite matters* (that is, those difficult to understand, perhaps even—from a child's perspective—beyond the reach of ordinary comprehension)? One has to be a specialist to operate the room (barbarians don't know how): it is the substance of an entire apprenticeship. There are those who would cite this knowledge under the rubric: *rule.* Certainly the behavior is regular that conforms to this knowledge, but everything we have written compels us to believe that it depends far more on the mutual acceptance of common goals (in this case, comfort) than on a mechanical application of rules which, as we have demonstrated, in any case are incapable of anticipating every contingency. Such an approach toward the rule, which makes possible the construction of robots, seems to us entirely to miss the point. The living of the room is not a following of rules (which in any event it generates) but the unfolding of a system of intentions within constraints which Experience (that of the Woods and all those from whom they have inherited and learned) has shown to be (generally [but far from always]) advisable.

It is these *constraints* which the Woods in common with most Americans call rules. As for the rest of it—the struggle toward a living *beyond the constraints*—is it not a connoisseurship?

How different is this from the more pervasive understanding of what constitutes connoisseurship: *expertness in matters of taste* (or discrimination); *knowledgeability, especially in aesthetic* (or recondite) *matters?* Only in the domain to which it is applied, on the one hand to the *appreciation* of drawing and collage, painting and photo; on the other to the *operation* of window and chair, table and stair.

But is the appreciation of a painting anything other than the way a painting is operated? (You operate a window by raising and lowering its sashes, by looking, making faces through it. You operate a painting by looking at it, wondering what it's about, making it the subject of discussion.) For the child commencing his living of the room, everything in it is equally taken for granted, equally inaccessible. The paintings are not the object of a catechism withheld from the stair (15); the chair (29) is not the subject of a lecture not delivered on the Lacandon drum (39). It is not that everything in the room is the same—the floor (8) is walked on, the rocker (35) rocked in, the records (27) listened to, the drawing (37) looked at (and frequently compared to its subject)—but as the child *assumes* the room, he takes on everything all at once; and if not in equal degree, then this inequality nonetheless does not observe the cleavage of the virgule in *aesthetic/utilitarian.*

How could it be otherwise with Mies van der Rohe's lounge chair (53), which, whatever its nominal status as utilitarian object nonetheless fully meets Panofsky's definition of art as that which "demands to be experienced aesthetically"? But the same is true with everything in the room, not because of any emanation (ours is neither an essentialist nor charismatic understanding of art), *nor* because of the intentions of its producer or Denis or Ingrid (who have transformed some utilitarian copper wire [63] into an aesthetic object simply by placing it in the center of the mantle [62]), but because this distinction is one of the things *learned* through the operation of the room in the unfolding of its living.

It is the difference in operation that *signifies* in the difference between *aesthetic* and *utilitarian*—which is why ironing clothes on the *Mona Lisa* is so transgressive. But it is the difference in operation that *signifies* in every distinction (*male/female, front/back, living room/kitchen*); and while, in a household with paintings and chairs, photographs and stairs, distinctions are internalized, those that distinguish paintings and photographs from stairs and chairs *are not* internalized differently from those that distinguish stairs and chairs. Put another way, just as one learns to sit up or slouch on the couch, so one learns (or does not) the pure gaze of aesthetic detachment (this helps explain the casual, often insolent, familiarity with legitimate culture so characteristic of those who have *grown up with it:* they appreciate its difference from their couch, but couch and culture stand from them at an identical distance [situating rather precisely their social class]).

Though in the following remarks from his *Distinction: A Social Critique of the Judgment of Taste,* Pierre Bourdieu is referring to the mastery of legitimate culture (painting, music, books), he could be speaking for us about the mastery of *anything* in the room (of anything in *any* room [so that connoisseurship comes to characterize a relationship to the world—that of growing up in it—not a relationship exclusively to the world of legitimate culture]):

> The competence of the "connoisseur," an unconscious mastery of the instruments of appropriation which derives from slow familarization and is the basis of familiarity with works, is an "art," a practical mastery which, like an art of thinking or an art of living, cannot be transmitted solely by precept or prescription. Learning it presupposes the equivalent of the prolonged contact between disciple [child] and master [parent] in a traditional education [growing up], i.e., repeated contact with cultural works [storm windows, chairs] and cultured people [parents, other adults]. And just as the apprentice or disciple can unconsciously acquire the rules of the art [the rules of the room] including those not consciously known to the master himself, by means of a self-abandonment, excluding analysis and selection of the elements of the exemplary conduct [becoming a parent's child], so too the art-lover, in a sense surrendering himself to the work, can internalize its principles of construction [meanings, values, affordances], without these ever being brought to his consciousness and formulated or forumulable as such; and this is what makes all the difference between the theory of art [sociology, developmental psychology] and the experience of the connois-

seur [the kid growing up in the room], who is generally incapable of stating his principles of judgments ["it's just the way we do things around here"].

Rules? Of course. One does not come to grips with a painting by tossing it on the fire, any more than one operates a storm window by putting a fist through it. But within the space carved out by the rules: a mastery of possibilities (acquired through slow familiarization) which can only go by the name . . . *connoisseurship.*

Mies van der Rohe's Lounge Chair

To make a chair as a falling leaf . . .

First of all, no legs. Then, a cantilever of stainless steel polished to a mirror finish: in the play of light, it disappears. Finally, two slings of a leather the color of humus: apparently weightless, they bounce on the spring of the cantilever like leaves caught in a capful of wind.

To a tubular-steel frame, Ludvig Mies van der Rohe has laced with rawhide two slings of saddle leather. The frame is a 1931 adaptation of his 1927 side chair, a somewhat higher, narrower, and straight-backed seat meant for dining. It is this side chair Mies van der Rohe insisted was "the first to have exploited consistently the spring quality of steel tubes." Not that the idea wasn't in the air. In 1885—the year before Mies van der Rohe was born—Max and Reinhard Mannesmann published their process for producing seamless stainless steel tubing (an invention that was to impress Thomas Edison more than anything else he was to see at the Chicago World's Columbian Exposition of 1893). The application to chairs must have been obvious. Steel tubing and/or cantilevers showed up in American designs of 1904 and 1922. In 1925, inspired by the steel tubing of bicycle handlebars, Marcel Breuer created the first version of his Wassily chair (along with a side chair, nested stools, theater seating). In 1926 Mart Stam made a cantilevered side chair out of tubular steel (coming up with an improved version a year later). It was Mies van der Rohe, though, who first exploited the resilience of thin-walled steel tubing to put a bounce in the cantilever. This allowed him to recapture the comfort of traditional upholstered furniture (with its buried springs) but in the modern idiom of exposed materials and revealed structure.

Though not all *that* revealed. The frame consists of two bent tubes of seamless stainless steel polished to a mirror finish. They are connected by dowels and screws hidden beneath the leather sling. Two stiffening bars, all but invisible below the seat, give additional support to the sling, held in tension by an all but invisible rawhide lace.

Another lace behind the back holds the back in tension. Two brackets welded to the frame make sure it doesn't slip, even when the lace is loose.

Does this make the chair dishonest? Candor is a subtle thing, and every revelation's relative (what's more hidden than the source of gravity?): the mechanics of the falling leaf . . . only *seem* to be revealed.

rules   196. "Don't sit down on it real hard" (Randall, kAPP, ktCON, tPRO).

197. "Don't bounce on it" (Randall, kAPP, ktCON, tPRO).

198. "Don't hang on back" (Chandler, kAPP, tPRO).

199. "Don't stand on it" (Randall, ktPRO).

200. "Don't untie the leather" (Chandler, ktCON, tPRO).

We have just seen that this chair has the nominal status of a utilitarian object while nonetheless meeting Panofsky's definition of art as that which "demands to be experienced aesthetically." At the same time it is a political tract.

At the very least it is a slap in the face. The oppositions are familiar. By revealing its structure, the chair unmasks the hidden: this excites because it is forbidden (it is eating the apple of the Tree of Knowledge). The chair is metal: this means that it is not upholstered, that is, not traditional (that is, deviant, Bohemian, unorthodox). The chair is air: this is to say that it has no burgher pretensions, that it has no legs.

Why no legs? Because this abolishes the chair as throne, archetype and source of all that is Paternal (the Voice of Orthodoxy speaks from the throne). Moreover, as Ludvig Claeser reminds us, Gropius, Mies van der Rohe, and Le Corbusier matured in bourgeois interiors which, from the perspective of the child, "must have appeared like a rain forest: innumerable richly machine-carved legs of pseudo-Renaissance chairs and tables." To bring down the Father: *cut the legs out from under him!* (Fraser would have understood.)

But *each* of these gestures is an attack on the priest-king (the transgression of Adam in the garden, the refusal of the patrimony). From this point of view the chair is little more than the reactionary product of a classic conflict between the generations (it recalls to us the reproduction of the room [no surprise that Denis's father put Saarinen's *pedestal* coffee table in the center of *his* living room]). Yet the very signs mobilized for this conflict point at the same time toward a utopian reconciliation of Mies van der Rohe's elitist aesthetic disposition (as part of the dominated fraction of the dominant class, what other disposition could he have?) with his socialist political inclinations (the universal dissemination of a revolutionary way of life), a fusion achieved in the *mass production* of *elegant modern* furniture. Given its instantaneous recuperation by the bourgeoisie and its widespread deployment as a mark of means (redoubled in the context of the table [65]), it is important to acknowledge that at the time of its introduction the furniture was in fact *affordable:* it really was industrial production for the masses, not the hand-crafted decoration for corporation and museum lobbies it has been turned into (*diversion of the revolution*). How serious was the political gesture it represented? Serious enough to be censured by the Nazis (the furniture was deviant, cool, and airy [it rejected the Father {hence: the Father-land}]).

Its allies in the room help to restore a little of its revolutionary luster. No one asks, "Is it really a chair?" as they do of David Rowland's (29), but they do say—in great surprise—"Oh! This is really comfortable." Yet the chairs are closely related. Both achieve a similar comfort in a steel tube frame, the one by *mimicking* that provided by the springs buried in traditionally upholstered chairs, the other by bringing *them to the surface.* Both use materials novel in chairs at the time of their creation (the steel tube, the polyvinyl choloride elastomer). Both are indebted to Thonet, for inspiration, for production (Thonet was the first to mass-produce Mies van der Rohe's designs). Both are tied, through the theme of bent materials, to the popular anonymity of the wicker rocker (35). The fact that the Sof-Tech chair is in mass production (airport lobbies, convention centers) flushes this latency in the lounge chair from its covert of saddle leather, mirror-finish stainless steel: the chair appears under the sign of . . . product design. By the same operation—but in the opposite direction—the luxe of the lounge chair (even to lounge is a luxury) deprives the Sof-Tech chair of some of the scandal of being a gangable, stackable chair in a bourgeois living room: suddenly it appears beneath the banner . . . *furniture: by/for architects.*

Which Gerrit Rietveld would insist was the point: "But as a piece of furniture a chair has other purposes than to look or be comfortable or 'not comfortable.' It, like other furniture, should contribute to making the space in a room tangible, to creating interior space—interior design as sense perception of space, color, etc." The whole issue of bourgeois comfort is, not dismissed, but . . . *put in its place.*

Utilitarian object, work of art, political tract, the chair is also an element in the room as . . . *articulated space.*

The room? Piece of sculpture.

The Right Speaker

Suddenly, in the quiet room—often startlingly loud—a fragment of speech from the speakers, "way downtown to see about" or "boulevard heading out to make," the volume rising and falling as the citizens' band radio broadcasting the voice approaches,

then recedes from the house. Or it jams the music, right in the middle of a Dennis Brain cadenza . . .

This is radio frequency interference, picked up by some element of the phonograph—the leads from the cartridge to the preamp?—and pumped into the room by the speakers, left (38) and right, these indistinguishable except for a tiny chip in the veneer at the upper right of the left side of the right speaker.

Radio frequency interference: it demonstrates the supreme indifference of these speakers to what they play: garbage in, garbage all over the room . . .

hand-off   6. "Same as other speakers" (Chandler, to RULES 164, 165, 167, 168, 169, 170, 171, 172).

The function of the speakers is certain (they are the device in the room responsible for transforming the quarter ton of vinyl chloride in the record cabinets . . . into words, into music). But what do they mean? The ear immediately hears in them the records (27); but the eye, observing their walnut immobility, seizes them with the same gaze that takes in the record cabinets (23, 26), couch (30), round lamp (42), and table (65). The mind reflects on the phonograph (amp, preamp, turntable), ultimately on the electricity that powers them, that is, on modern technology. Song, wood, electric current: they sketch a certain history of mankind (it is one audible in the records stored in the cabinets [chant of Gregory, horns of Mozart, electrical sonance of Varèse, Cage, Crumb]). It is also one that erupts in the room, where the technology of the speakers usurps the voice of the family.

This is not a family that sings (what contemporary bourgeois family does?). On birthdays there is "Happy Birthday," at Christmas caroling in the neighborhood (but even then the kids are embarrassed, come along, but do not sing), and little else (the stuttered accompaniment to records, tapes, of those who . . . "sing along"). The musical life of the family—evidently important (vast resources are devoted to it)—is reduced to *choosing*. Here Barthes is wrong. The advent of the technician does not abolish in the musical order the notion of praxis, just *dislocates* it: in place of the desire to *make* music comes the desire to *own* it. This is subject to a profound economy; and therefore the *playing* of music will be replaced by a *shopping*. An abundance of catalogues will display a cornucopia of the world's music past and present: these will be studied, annotated, filed. Chains of record stores will mount bins, racks, down aisles disappearing in the distance. Each is crammed with records, tapes, discs: these will be flipped, pulled, noted. Records will be reviewed. Whole journals will be devoted to nothing else. With age there will mature markets for used records, tapes, discs. These will circulate like collectable coins, no longer tokens of *another* value (that of music, money), but their *own:* "mint," "very good," and "good" will refer—as with coins (which like records are exactly: pressed)—solely to the token itself (quality of the vinyl, the pressing, character of the inner sleeve [must have the right advertisements], condition of the jacket [a secondary market will develop for these alone]). Hence the phenomenon well-known to buyers of previously owned records: they have often . . . never been played.

What is missing here? From this commodities circus—stars, superstars, arrangers, jacket designers, contracts, performing rights, plagarism suits, conductors—the music has . . . disappeared. When it returns it will no longer be played, but listened to. That is, music will assume exclusively the posture of its consumerist appropriation (Michael Jackson, the Canadian Brass, Luciano Pavarotti).

Is it a coincidence that this has happened in our room? Hardly. The implicit alienation is precisely that embodied by a living room scarred by the violence of a stair (15), central shrine in the church of the single-family private home (tombstone on the grave of the face-to-face community). The speakers strip music of the contagion of community implied in its production (harmonizing with others, following a conductor [submergence of the ego in communal effort]) at the same time that they tame it. Is it too loud? Turn it down. Too irritating? Turn it off. Too threatening? *Throw the record away* (say: *it's just bad music* [it may be, but in any event, blame the Other]). The music that ends up being listened to corresponds to a very narrow band (hence the proliferation of radio stations playing *your* favorite music, each precisely tuned to an ever constricting annulus); predictably, this produces increased possibilities for contestation (but none for solidarity), each exploited by one or another marginal fraction of the sum (adolescent, intellectual, black, redneck, college student [adolescent intellectual, redneck college student, black intellectual college student {and so on}]). Necessarily, *all of them* will oppose in one way or another the individualistic and competitive values of the bourgeoisie (the center, distance from which will define and confirm their marginality). Therefore, should such bourgeois values advocate the *nuclear family,* music will speak for the *tribe* (notoriously, Woodstock, but marches are not family music either, little dance music is, musical theater implies the company, religious music the church, even most intimate music is not of or for the family [John Dowland, Thomas Tompkins]: what's left? Mitch Miller sing-alongs). If such values include *sexual discipline,* music will embrace *sexual abandon* (from Dowland's "Come Again! Sweet Love Doth Now Invite" to the Beatles' "Why don't we do it in the road?"). Should the bourgeoisie admire *material ambition,* music will exalt the *dreamy,* the *ecstatic* (from the sequences and hymns of Hildegard of Bingen to Marion Williams singing "They Led My Lord Away," from Debussy's *L'Après-midi d'un faune* to Crumb's *The Ancient Voices of Children*). If it is *measurable accomplishment* that is at stake, music will be dismissed as a *waste of time* (the myth so many musical biographies cite as the origin of every inspiration).

Worthless, dreamy, ecstatic, abandoned—no wonder the music has to be sealed in plastic (and it is possible to make a buck in this too). In the great market of vinyl and cardboard, polyethylene and ink that results, there *is* a kind of human intercourse (though even this is debased by the record chains whose employees know nothing of music and who in any case quit as soon as they can). It is not, however, the intercourse envisioned in or by the music. Sanitized by its packaging, strangled by the speakers, all this music is, even if heard . . . diminished, muted, reduced to a sign of the room's plural, of Denis and Ingrid's means, of their connoisseurship.

"Den," Ingrid calls from upstairs where she and the kids are getting ready for sleep, "could you turn that down a little?"

Of course he could.

On the couch in the living room with Mahler's *Symphony of a Thousand* roaring from the speakers, in command of the full resources of a huge orchestra, two-part chorus, boys' chorus, and eight solo voices, immersed in the cosmicizing universalism of the anthemic plea, "Infunde amorem cordibus," Denis nevertheless is alone, suspended between the speakers and the stair in a privacy he neither enjoys nor knows how to live without.

Is this self-pity? Perhaps. But among institutions of privatization speakers and stair stand tall. And the couch does slouch between them.

The King

It is in the soft light of early morning that the king is most eloquent. Then the terra cotta is animated into flesh, the incised beard quivers with excitement, the carefully modeled mouth trembles on the verge of singing.

It is a foot high, from crown to base, where ridiculously tiny feet peak from the hem of a robe better suited to the *Balzac* of Rodin. Open at front, this reveals a garment covered with the appliqué for which Teodora Blanco is notorious. Twenty-six vermiform squiggles squirm across this surface which as well is spotted with a number of dots. The red slip of gown and crown is also daubed here (it is burnished on the gown, not on the crown). Cup in hand, cross at throat, the figure is touched by a transcendent detail: a splendid modeling of the collar.

Conventionally characterized as a king, of whom? of what? Perhaps of no more than the dawn . . .

rule   201. "Not supposed to play or mess around with the king" (Randall, tpro).

What are the squiggles and dots?

Could they be other than the mark of the jungle of dreams where scarlet birds rule (39)? Authorized by the delirious zoomorphs inhabiting the rest of Teodora Blanco's work, we cannot but believe they are. In this way the made-for-sale pottery of this surrealist folk sculptor from Santa Maria Atzompa is united with not only the Lacandon drum (39)—with which it is is doubly bound as pottery and by a shared history of purchase in 1961—but the drawing of Ron Rozzelle (39) and the lithograph of Max Ernst (20). As "primitive" figuration, it is linked to the paintings of Peter (41, 64); as folk-craft made for sale, it joins the Cherokee basket (24), Ferris wheel (25), throw pillows (36), wooden car (60)—with three of which it is further linked by a common origin in Oaxaca—and the baseball game (68), to all of which it is further tied as a souvenir of a consumerist travel to foreign shores. At Christmas a sprig of holly turns it into seasonal decor.

Effortlessly the signifier slithers free . . . of every chain.

The Paul Strand Photograph

On the white wall above the king, above the black grill cloth of the speaker, behind a sheet of plastic, within a steel-gray frame, in a white mat, on an eggshell ground, inside a black neat line, a mess of grays . . .

This is Paul Strand's "Photograph, New York," a photogravure on Japanese tissue from the June 1917, issue of *Camera Work* (no. 49/50), "A Photographic Quarterly. Edited and Published by Alfred Stieglitz. New York." Because of its subject matter, it is often referred to as "Sandwich Man."

The frame is another Kulicke, a slip of aluminum, welded at the corners, doing its best to meld with the wall.

hand-off   2. "Same as every painting" (Chandler, Randall, to RULES 90, 91, 145).

Who is this man?

We cannot say. We *can* say that Strand took this photograph in the first flush of an excitement inspired by Lewis Hine, his teacher at the Ethical Culture School (at the time Hine was commencing *his* documentation—compassionate and unflinching—of immigrants, poor at home and at work, laboring children, devastations of war). Strand's photograph—as Hine himself—is steeped in a radical politics of the left, in a visionary egalitarian democracy, in a Whitmanesque transcendentalism (connections

to Peter's box [31], paintings [41, 64]). Yet after this has been said, *who is this man?*

To insist on his stubborn "factuality"—on his realism—is to overlook the inescapable *idealism* that is consequent on: the exclusion of any context, the way he fills the frame (he overflows it), the way the aestheticism of the photograph (it is a nocturne) forces us to confront in him his form, to acknowledge that this is, if not God-given, then Natural, obliging us thereby to discover the Social forces which would "cloak" it (as the signboard does, literally and symbolically). The formal purity of the image is thus warped by its subject into an ethical purity (the reverse is also true). Here an unsuspected connection to the lounge chair (53) and coffee table (65), whose formal purity is as well an analogon of moral purity: "I don't want to be interesting," Mies van der Rohe once said, "I want to be good."

(The Kulicke frames, here, on Peter's *Malatesta* [41], on the Yunkers *L'Amoureux* [40], express a similar purity [it is Puritan, it is almost Shaker].)

How much of this do the kids appreciate? Very little. Questioned at nine and eleven, they saw "a man who's probably poor doing street advertising," but assumed that Tom Koch had taken the picture (Tom, a family friend, has a couple of pictures in the hallway upstairs along with another Strand, Walker Evanses, and Eggbert Hansens). "Maybe it's in New York," Randall says, but he turns out to mean Buffalo, "Because Tom lives there." (Actually, Tom grew up in Buffalo, though at the time Randall said this, he was also a frequent visitor, caring for his declining father.) Because of associations with Ingrid's father, who spent his last years in Charlotte, Chandler thinks it might be set in there. Randall observes that there *is* an 18th Street there. Why does the photograph hang in the room? "To amuse," says Chandler. Randall says, "Only to beautify it."

There is, of course, *no* pure reading. Ingrid remembers Denis's father dividing the photographs among the boys, Christopher getting "White Fence," Denis "Sandwich Man," Peter—but who remembers? (He sold it anyhow.) For Denis, this goes further. His father was a photographer in the tradition of Hine and Strand, working among the Amish, Cleveland blacks, in Mexico. One of his photographs was in *The Family of Man,* the blockbuster exhibition of the 1950s organized by Edward Steichen who, in the days when he spelled his name Eduard, had been a mainstay of Alfred Stieglitz. There is thus a kind of thread that runs from the photograph on the wall through Stieglitz and Steichen to connect Paul Strand and Denis's father. This lineage never precluded the articulation, however, of a fundamental difference between Steichen and Strand. Less defined than exemplified in other domains (structuralist paradise!), this was paradigmatic of a whole set of distinctions that constituted a kind of Woodian aesthetic opposing *Steichen* in photography (/ *Strand, Hine, Evans*), *Russell Wright* in design (/ *Charles Eames, Eero Saarinen*), *Carl Sandburg* in poetry (/ *Kenneth Patchen*),

Norman Rockwell in painting (/ *surrealists, abstract expressionists*), *The Saturday Evening Post* in literature (/ *Henry Miller*), *Elia Kazan* in film (/ *Luis Buñuel, Jean Vigo*). The aesthetic, barely limned here, was comprehensive, subtle, concrete (as little kids, Denis and his brothers were more at ease humming the theme to Vigo's *L'Atalante* than any patriotic hymn). It also constitutes in Denis and Ingrid's living room that Voice in which we have detected a dead modernity, dead because no longer fructifying in a world where—betrayed by those who admired without understanding—each of its contestations has been recuperated as a "classic," and thereby drained of threat and promise, energy and life. The century has not turned out the way its best imagined it might. The dizziness is gone. All that's left is a frail and fading beauty.

The room? Requiem, elegy . . .

The Fireplace

As the sun reddens, so the leaves across the street. Trees once a torch of yellow cool to brown. In the afternoon, low light across the porch, through the windows, along the floor: sunlight on the hearthstone!

On the hearth *brick* that should be, for the hearth is brick outside the fireplace, concrete within. This latter is nearly three square feet in area, paving a firebox volume two and a half feet high run up in fire brick. This vanishes into the throat (stuffed with newspaper) and damper (operated by a chain), beyond which the smoke chamber narrows into a flue (lined in terra cotta) that rises up through the chimney (no fewer than thirty feet). Presumably all this works, though since Denis and Ingrid have never set a fire in it, it's hard to say. The firebox is framed by forty-three fireplace bricks whose rounded corners smooth the transition from the mantelpiece. This consists of a fascia nine inches by forty-one flanked by pilasters four feet high that terminate in molded plinth blocks. Simple brackets protruding from the pilasters apparently carry a mantelshelf ten inches deep, five inches thick, and five feet long. An outer frame composed of two vertical fillets surmounted by a baseboard molding completes the ensemble.

And when the sun touches its white enamel coat, it flares like a Pharos!

rules   202. "We shouldn't play in it" (Chandler, kAPP).
203. "Don't pull the chain" (Chandler, Randall, ktAPP, tCON).

The fireplace is a fraud: it is *not* the place of the fire.

The fire has been banished: to the basement (heat), kitchen (cooking), lamps (light). Why then in the living room this medieval contraption of bricks and boards (a

hundred and sixty-eight bricks that we can see) locking up thirty-two square feet of wall, pushing out almost two feet *into the room*, a foot and a half back *into the wall*, which had to be rotated forward—sweeping out ten square feet of floor space—to accommodate this protrusion into what otherwise would have been a membrane a handspan thick between living and dining rooms? Why this brick tail pipe coughing smoke into the skies? Why the risk so wantonly invited by the open flame, the cracks in the flue, the sparks on the roof top?

It is for the god who dances in the living flame, they say (Gaston Bachelard says so and Christopher Alexander, Lord Raglan and James George Frazer), but what god was ever worshipped in his essence alone? Before the reverie (or with it: we have no pragmatist's axe to grind) came warmth and cooked food and a light to conjure by. The exile of the god to the basement *may* be our loss (though nothing new for him, Helios every night descending into darkness, Baldur going down to Hell), but the pretense that anything numinous could endure in the *occasional* fire is worse—the occasional fire ("wouldn't a fire be nice?"), the gas logs, the tinfoil coals (with their motor flickering a light [the cord sneaking away from the fireplace {to the real god at the powerhouse}]). Worst of all is the empty hearth: this is to *enshrine* emptiness (it is genuinely Satanic). A god without a role is diminished, is become like Nature in the room, a dumb show, a mime (here the mime is of Tradition [why not have a diorama just inside the door: The Good Old Days?]). None of which is to deny the hold of fire on our psyche or the pleasures (though we may doubt them) of chestnuts roasting on an open fire. It is merely to observe that with their superfluity they become the pleasures of a certain class (après-ski), of a certain exploitation.

It is precisely this superfluity that the bourgeoisie find so attractive. Nothing so effectively marks their difference from those with fewer resources and scarcer leisure than a devotion to the useless; and in a house which heats with oil, cooks with gas (and in any event not in the living room), and lights with electricity, a fireplace is useless-ness cubed. At the same time, nothing is less easy to acknowledge. From the perspective of bourgeois utilitarianism, cults of the useless are perverse (they are: aristocratic). Bourgeois pragmatism, common sense rebel. Therefore the fireplace *will* find a role: endless mumbo-jumbo (invocation of ancient gods, household deities [the glory that was Greece, et cetera]), psychologizing, philosophizing (here is Bachelard: "to be deprived of a reverie before a burning fire is to lose the first use and the truly human use of fire"), hand-waving (here is Christopher Alexander: "the need for fire is almost as fundamental as the need for water"). Against its notorious inefficiencies (two-thirds of the heat goes up the chimney) a pragmatic will be nevertheless advanced (here is Time-Life's Home Repair and Improvement series: "still, a properly main-tained and well-fueled fireplace can be a useful auxiliary to a central heating system").

In *A Pattern Language,* Christopher Alexander even manages to admire its utility as babysitter (and this under the banner of "down-to-earth"). Here he is quoting Robert Woods Kennedy quoting Mrs. Fields:

> During the winter months, when the children are often confined indoors for their play, it often happens that around four o-clock or a little after they become cross and grumpy in their playroom, or wild and almost hysterical with boredom. Then I light a fire in the living room fireplace, and send the children in there to watch it; if the fire were not lighted they would continue their quarreling and perhaps try to turn the quiet room into another bedlam, but with the burning flames on the hearth, they relax into easy interest.

What do we hear in all of this? It is the voice of a bourgeoisie that doesn't even know it's speaking. The rhyme here is with the stair (15): the dangers the fireplace imposes on the family are the risks *the house* accepts in order to complete the syntagm of classical patriarchal Western culture supporting and supported by wealth and power (a two-story single-family house with a fireplace in the living room). The fireplace joins the Georgian Revival door (2, 3, 4, 5) (with its incorporation of Rumford's eighteenth-century modifications to throat and firebox the design of the fireplace is Georgian in fact), the molding around the window by the stair (17), the newel post (10) and the rest of the woodwork (11, 47, 50) in the desert of pretension that the syntagm of modernity (6, 16, 20, 21, 23, 24, 25, 26, 27, 29, 31, 32, 33, 35, 37, 39, 40, 41, 42, 53, 55, 56, 60, 63, 64, 65) struggles to irrigate with the irritation of a vital living. The confrontation is one we have already seen: *what are the useless flames but wallpaper?* In the name of the clean, the efficient, and the essential, modernism rejects this fireplace which the upwardly mobile petit-bourgeoisie brandish in the name of a bankrupt Tradition (nothing so clearly unmasks the revanchist program of postmodernism as its reclamation of the fireplace [it has even discovered an unsuspected taste for gas logs, tinfoil coals]). In resolving the opposition of a genuine *utility* and a class-conscious *display,* the *house* is the expression of a culture (again, Denis's father's father's) that values show above use (what it displays is its taste for the things of the class above it [it is, precisely, a dominated aesthetic]). Modernism, confronting a not dissimilar opposition, arrives at a not dissimilar destination (what it displays is its taste for the necessary, the essential, the dizzying [as the taste of the dominated fraction of the dominant class, it attempts to *vacate* the marks of differences in means]). Linked through ties of reciprocal definition, in the room these stand, mutually critical, as a Protestant to Catholic, united in their vision of *an* order, divided in their vision of what that order ought to be (at stake are two utopias). Their debate—acrimonious at points (as when the white walls [16] take on the white door [2, 3, 4, 5]), harmoniously sweet at others (they pretty much agree about the floor [8])—renders the room ecumenical in tone (an ecumenicism amplified by the saturation of certain paradigms [for example, *folk/popular/elite*]). Yet this very ecumenicism is not without significance in the debate: in that it rejects the hegemonic instincts of the house, it claims the room . . . in the name of the modern (in whose name: no fire in the fireplace [furthermore, having painted the room, Denis was in no mind to see it painted again {in smoke}]).

Does this mean Denis and Ingrid have no appreciation for the charms of the dancing god? Not at all (they too are bourgeois). It only means, in resolving the opposition of petit-bourgeois pretension (of the house) and bourgeois promise (of the modern), that they make of the room *the expression of a liberality.* They accept the fireplace . . . *as sculpture.*

The room? Art gallery.

XXXII · THE APPEARANCE CODE

The appearance code wants to be taken at its face. But we know better: *nothing is to be taken at its face.*

Don't slam it? We understand immediately this not a question of using or protecting the door, but whether the kid letting it slam exhibits any manners. *Don't hurt the plants?* Here the life of the plant is at stake, but not more so than the ability of parents to raise kids who know better. *Don't breathe on it?* Instantly we recognize the role of the glass and its cleanliness in the social and moral economies of the home. Always less important than the smudge on the glass, the health of the plants, the noise of the door, is the appearance of the kids, not their color or shape (in any event a sign of their health), but their personality, their character (are they barbarians? are they civilized?). Without reflection this in turn is referred to the parents: are Denis and Ingrid fit? Do they have it when and where it counts? The issue is depth, depth of civility, depth of character (through and through? or lying lightly on the surface?): the face is only an index to the soul, the body to the mind. Form informs content, effect speaks to cause, appearance to reality. But where does this reality lie?

On the surface.

Where else? We know the world only through our senses. To be (perceived) culture is obligated to be ostensive (the "depth" of culture is not an essence, some "inner" culture, but biology [skin, tissue, DNA]). Culture manifests itself (which is its mode of being) as a system of ostents: the absence of the smudge makes a point, the thriving plant makes a point, the absence of noise makes a point. And what is this point? It is that culture is present. And in what does this culture subsist? In the *not* smudging the glass, the *not* hurting the plants, the *not* slamming the door; in, that is, the surface that is the sign of the presence of the culture that generates the surface. Culture is the making of signs that point to their having been made (anything else is natural). The clean is always the sign of the well-organized household. But what is the well-organized household? It is clean. Demystification of the code unmasks the essential tautology at the heart of every system of meaning: the clean is to be clean . . . to be clean. Culture appears as a sign endlessly referring to itself. It is a fistnote whose index finger points back to its own hand.

The emptiness implicit in the tautology is stuffed with the clutter of the rules. *Do this, don't do that,* there must be some purpose. It is less, thus, that the appearance code wants to be taken *at* its face, than *as* a face, that is, as a mask, an outward show, a semblance . . . of something else, something within, something Real. But there is no inner Real, behind the face of the appearance code there is no soul.

What you see is what there is.

The Plant in the Fireplace

It's a sort of buck-and-wing, up, up, straining, down, bouncing, elastic, up, up again, and down, again bouncing, and up—*look*—this leaf flipping, that waving, brave on their long spathed stems, green dancing to the rhythm of the fan . . .

A monstera dancing takes up a lot of room, thirty-five cubic feet in this case, about the volume Denis, Ingrid, and Randall would occupy if stacked like cordwood before the fireplace. The spadices spring from potting soil in a terracotta vessel nine inches across, nine deep in a terracotta saucer. This sits right on the bricks of the hearth.

The fingers of the deeply incised leaves seem to snap and pop, as if keeping time: cool stuff, that fan song!

hand-off   7. "Just like all the plants" (Chandler, Randall to RULES 84, 85, 86, 99, 181, 182, 183, 184).

Bursting from the fireplace, the monstera is aflame. Green tongues of fire lick the air, beat upon the room with each pass of the fan.

This is the clear meaning of the plant, to be the fire in a fireplace (57) from which fire has been banned. This accelerates the drift toward Nature of a plant already in accelerated motion, marked like that on the record cabinets just inside the door (22) by every variety of overheated excess (it is voluptuous, it is expansive, it is unrestrained). Fanned flames, tropical plant, it is an incendiary combination: in the middle of the room . . . *jungle fire,* archetype of all that is wild, undiscriminating, all-

58

consuming. It is Nature, *completely* out of control . . . on the very hearthstone of Culture (this is the horror at the heart of Nature [the return to the Chaos out of which god created order {including that of the social classes}]).

Not to worry (it is the lion tamer's head in the mouth of his beast): the fire is purely symbolic, the plant is in its pot. By intensifying the sign of the nature subdued, the authority of Culture is magnified (etymology: tame plants). Here, at its hearth, everything is really . . . *under control.*

Even the cricket is out of doors.

The Plaited Fan

Who was it who heard "poetry in motion" as "oh a tree in motion"? We disremember, but this is it, a serrated palm frond rewoven as a cordate ovate leaf like that of the lilac, hanging by its stem to the underside of the mantel, twisting, swaying, in the slightest breath of air.

This is the plaited fan, one of a handful Ingrid found downtown in a dime store to make a present for Denis's birthday the summer they moved into the house. A frond has been stripped into lengths of equal width and—still attached at the stem—plaited together into a fan. The returning ends have been wrapped against the stem with a piece of plastic raffia making a strong handle with a loop to hang it by. Into this Denis has threaded a length of clear monofilament fishing line by which the fan hangs from one of hooks that at Christmas have stockings depending from them.

Is it the palm that sways? Or the lilac? Poetry in either case . . .

The fan is a guest at two parties.

At the first it plays the camel bells (6) to the fireplace's (57) door (2, 3, 4, 5). Here it is part of the haze of signs that manages, despite the overweening presumption of the white and the bright, to produce in the room a diffuseness, an effervescence of meaning. Here twisted plait couterpoises itself to the smoothness of the mantel, the matte of the fond to the gloss of the enamel. The fan's cordiform perimeter mocks the mantelpiece's carpentered orthogonality. Its incessant motion makes light of its stolidity. The very marginality of the fan increases the force of its iconoclasm. This is not folk art on a pedestal (it is not the king [55], not the Lacandon drum [39]), but folk craft lightly embroidered onto the utilitarianism of the room, hanging from the mantel of a fireplace the sign of everything its Southeast Asian weaver's life is not. The fan is the sign of the room's doubt about the legitimacy of this difference, one echoed

everywhere, but with the fan especially in the Cherokee basket (24) and the wicker rocker (35).

Like every hesitation expressed by the room at the expense of the house, this is recuperable. At this second party the fan is an object of self-evident utility. However, this self-evidence is more conventional than real. In the first place, the fan is no less evidently decorative. It is not sitting at hand (for instance, on the table) but hanging from the mantel in a kind of wanton . . . *exhibitionism.* In the second place, who needs fans in the New South of Carrier and Trane? Given air conditioning, the fan's cooling function yields precedence to its sign function. It is made to speak . . . of another time (chair back, feet on the porch rail). In the presence of an active electric fan (61), this sign is complicated. It is made to speak of an even earlier time, one before electric power (there are two handcrafts embodied in the fan). But any nostalgia is immediately recuperated by the implicit criticism: while we sit idly (the fan limp on its hook) the electric fan passes to and fro (powered by the lives of the miners). The old days *were* better (less exploitative [and porch sitting encouraged a face-to-face life]). But then Willis Carrier brought us all indoors. Suddenly the meaning of the fan is certain: it marks the refusal of air conditioning on the part of the home.

But does this not contradict the Voice of Modernity so distinctly sounded in the room? Not at all. Modernism is not some promiscuous embrace of contemporary technology but the promulgation of a way of living free of inherited strictures, above all, those of social class. It looks toward an open commerce of peoples uninhibited by convention, whether of habit, clothing, or building design. But air conditioning is above all an agent of that privatization so strongly associated with the single-family private house (with its ally television, it insures an empty street). In the room this allies it with the white and the bright (with the mantel). Air conditioning is distinctly utopian, but its utopia is that of the bourgeois ego. The fan speaks for community, intercourse, solidarity (with respect to the mantel it is doubly iconoclastic).

The room? Hot . . . *but friendly.*

XXXIII · UNSPOKEN RULES

What are unspoken rules?

They should be the essence of connoisseurship (the connoisseur is the one who cannot specify his or her principles of judgment), but actually they do not exist (rules are spoken by definition). What Randall refers to when he speaks of rules "not written, not spoken" are what Ingrid refers to when she speaks of rules "totally unspecified." Both are talking about rules whose scope is relatively universal—that is, on the scale of the room, even the house. In our inventory the issue first arose with respect to Peter's box (31). Where Chandler volunteered *don't touch—just look at it* (RULE 145), a rule taking the box as its sole object, Ingrid came up with *keep hands off* and *don't throw things,* rules she insisted had general objects "totally unspecified." (Randall's "not written, not spoken" is in fact his gloss on her comments.) Later on, discussing the room in general, she put it like this:

rule   A. "Don't throw things in the living room—you might hit something" (Ingrid, kAPP, kCON, tPRO).

We've cited this under each of the codes for obvious reasons: a kid is he who does not throw things in the living room (kAPP), *and* what he does is throw things outside (kCON), though the real force here is to protect valuable things (moreover ones highly cathected) from accidental damage (Peter's box [31], the Lacandon drum [39], the king [55]). Because the fear is of *random* damage, the injunction is laid down *over the room* (and how else to deal with the problem? add a *don't throw things at* to every vulnerable object?). Out of this practice there may well develop an understanding (that of the connoisseur) about where and where not to throw things, but it will not be a rule (it will be vague, general, diffuse [it will be capable of addressing contingencies]).

Here's another example, one without an object of any kind (though not without an objective embracing a dozen objects):

rule   B. "Walk in the house!" (Ingrid, kCON).

This is as frequently heard as *no running in the house,* and in various splinters, one of which shows up as a stair rule (*no running on the stairs, it makes too much noise* [RULE 71]). This probably has as much justification to be cited under all three codes as *don't throw things in the living room—you might hit something,* for certainly one of its ends is protection of kids and things from the damage sure to result from high-speed, out-of-control careening-around (ktPRO), and another must be reduction of the sort of implicit animal abandon that in particular gets under Denis's skin (kAPP). But the rule is so unambiguously a rule specifying what a kid *does* as to be paradigmatic of control rules on the kid side (*no running* is another story: it *would be* tri-coded).

In general, though, the more general the rules become, the simpler their form (the narrower their coding) and the broader their force (everything is implied). For example:

rule   C. "Keep the house safe" (Ingrid, ktPRO).

Certainly the control code is implied (a kid does things that keep the house safe) and we need not explain how this encourages a kind of appearance (it would indeed be a concordance of blessing). Were there any doubt, though, about the primacy of the protection coding, Ingrid's gloss puts it to rest: "Keep spaces clear so no tripping, so firemen don't trip if they come in the night." This carries a twofold implication: one is without doubt the safety of the kids and house from death and destruction (this is the ultimate concern [in some manner it lies behind every rule {Death did not wait to enter the room through Peter's box: it was always there}]). But another materializes as a cascade of ever more particular rules directed against ever more limited forms of mayhem. Here we have dropped down a level: *no leaving things on the landing* (RULE 69). And here yet another: *don't leave stuff on it so people can trip, like marbles* (RULE 48).

There are even metarules. Doubtless there are many of these (from *there's no point to having a rule if you're not going to follow it* to *a rule you can't break's not worth having*) but we were able to solicit only one:

rule   D. "Explain to your friends what the rules of the house are" (Ingrid, kCON).

Here is a rule outfront about the collaboration expected from the kids if the home is to work. Only last week Denis obligated Chandler to reiterate to Kelly the rule about getting things from the refrigerator (he is to ask) after Denis had exhausted his authority by shouting at him (he kept interrupting Denis who was trying to read the newspaper on the couch). In the end Kelly left in a snit, not to return (while not a unique occurrence, it may be useful to understand that Kelly has been here for hours every day for weeks). By spreading the burden of the rules throughout the family, they are depersonalized. It cannot be a question of what (weird) Denis wants or (crazy) Ingrid when the kids too are spouting rules left and right. The rules come to be experienced instead as what profoundly they in fact are: *an aspect of the room*. Although it is Randall—in his finest collaborationist's garb—who intones:

rule   E. "Don't treat the room as a jungle gym" (Randall, kAPP, kCON, ktPRO),

it is heard as the materialized articulation of the room itself. "Please" it says—and everything about the room chimes in, its neatness and cleanness, its distinctiveness and order, its openness and whiteness, its lightness and color, the way Ingrid is in the room and Randall, Chandler, and Denis—"please, don't treat me as a jungle gym."

Only a barbarian could fail to hear this . . . "unspoken" rule.

The Wooden Car

"*Clack-clack-clack-clack-clack—*"

It is always the big wooden car the little kids are drawn to play with. "Please don't touch, dear," the parents say, praying inwardly that nothing's been broken.

It never is. Only the sun has ever hurt the car (though its license plate has vanished), the sun and Homer who shat on it the day Denis and Ingrid moved into the house. Homer was a dog who liked things to stay the way they were, and Denis had to repaint one of the wheels (but which? fourteen years later it's impossible to tell). The car is solid. Each of the forty-four pieces of wood is solid, some are substantial (*and* scavenged: the wheels were cut from a board from which the label was only incompletely removed). The base is a slab of three-quarter-inch pine and the rest of the car is built to match, twenty-two inches long, ten high, and ten wide. Eight pieces of galvanized metal make a rear bumper, running boards, and back and front fenders. Two others make the clacking sound, when the front crank is turned (clacker under the hood), and when the car is pushed along a resistant surface (clacker on the rear axle). The whole thing is held together with a hundred and five nails and a handful of staples.

There are four people in the car (one with hands on the wheel) all of whom could easily cram into one of the seats on the Ferris wheel (25). No surprise. Denis and Ingrid got the car from Nance and Jap, who bought it in Oaxaca from Enrique, who more than likely bought it from the family of Pedro Hernandez (who made the Ferris wheel). Whoever made the car painted it the same bright aniline colors, which now have faded to the same shades of pale, phantom and ghost.

"It's okay. He can play with it."

"Well," hesitantly, "if you say so—"

"*clack-clack-clack-clack-clack. . .*"

rule   204. "We shouldn't wind it up and play with it" (Chandler, Randall, tpro).

The car is more or less a toy. This despite a rule that says: *don't play with it.* What can this mean?

That it is neither a trophy nor an objet d'art, or, if these, then less so than either Ferris wheel (25) or king (55) and so, *relatively* a toy. The Ferris wheel *is* a trophy. This is attested to both by its extraordinary size (it is a trophy *in the form* of a toy) and the consequent complications of its removal (it had to be crated, trucked, there was an intercession of . . . forwarding agents). The king *is* an objet d'art. It suffices to point to its site to demonstrate this: it stands on a pedestal (54). The car *is* oversized, but not enough to bespeak crates, cranes, the lowering into cargo holds: it is not a trophy. The car *is* spotlighted, but on the very floor where it would be . . . as a toy. Furthermore, *it is played with,* if only now and then (Denis, the kids, show littler kids how it works: they play with it). Does this mean the wooden car is not a trophy, not an objet d'art? Not completely. It means no more than that it is one whose function as a toy has not been entirely usurped.

Although the car shares none of the connections made by the king (55) to the Ernst (20), Rozzelle (37), and Lacandon drum (39), it more than makes up for them by tapping into the networks of populist inclinations radiating from the Ferris wheel (25)—that is, to the refusal of the lock (7), Peter's paintings (41, 64), the populism implicit in the Voice of a Dead Modernity. Other connections: with certain records (27), the couch (30), Yunkers (40), and Strand (56) as previously in the possession of Denis's parents; with the Cherokee basket (24), Ferris wheel (25), throw pillows (36), Lacandon drum (39), king (55), and baseball game (68) as folk art; with the Cherokee basket (24), Ferris wheel (25), and king (55) as folk art made by a named individual; with the Ferris wheel (25), one of the throw pillows (36), and king (55) as originating in Oaxaca; with the Ferris wheel (25) as an object of poignancy fading in the sun . . .

XXXIV · THE PROTECTION CODE

Nothing could be more straightforward than the protection code.

This is because its function is so simple (so *rulelike* we could say). It does not assume the burden of *defining* culture (this it entrusts to the appearance code). It does not accept the labor of *reproducing* culture (this job it leaves for the control code). All it does is . . . protect culture, defend it, keep it from being scratched, broken, smashed, knicked, mangled, shattered, cracked, chipped, weakened, ruined, marred, worn, botched, blighted, scarred, torn. It's a kind of a brine (vinegar, formaldehyde). It achieves its end by interdicting actions for their own sake, not because of any light they might shed on the animality of the kids or the incompetence of the parents: *don't push, don't kick, don't hang, don't sit, don't slide, don't jump, don't stomp, don't touch, don't turn, don't play, don't pull, don't do anything* (it really is an embalming fluid).

Why is its language so negative (it scarcely knows any other form)? Because it is less concerned with *shaping* behavior (the role of the appearance and control codes) than *delimiting* behavior's scope, even when we grant that this is the role of *all the rules* (seven of every eight protection code rules are *don'ts*). This is not to say that the code embraces no positively modeled rules: *cover speaker if you're doing work around it that might allow stuff to fall on it* (RULE 98) and *make sure you put sleeves on records so that opening is up* (RULE 116) are examples of rules emitted *to protect* (speaker, records), which nonetheless manage to model behavior at the same time. But they also demonstrate the limitations that inevitably characterize rules of this orientation: they are prolix, unmemorable, and incapable of addressing contingencies (there are a couple of rules which don't fit this description—*be careful* [RULE 164] and *be VERY careful* [RULES 103, 106]—but these verge on vacuity).

As we have repeatedly seen, it is easier and more effective to keep cattle from the garden with a hedge (of don'ts). And what are the kids if not . . . calves?

XXXV · THE CIRCLE OF THE CODES

Round and round the circle of the codes: the room is to be neat so that we can keep it neat. *Here:* the room is *to be neat* (what is this but the phenomenal surface ruled by the appearance code?) *so that we can* (what is this but the form of a telos shaped by the control code) *keep it* (what is this but the ontic state dreamed by the protection code?) *neat* (what is this but the return of the appearance code?). Round and round (it is the dance of culture): the sentence slips like a rosary from hand to hand, but it is not the sentence alone that does this, it is the rules, which in their shimmering circuit from code to code (from tPRO to ktAPP to tCON to kPRO [we have watched them do this]) slip ever . . . *just beyond* (to the next code [which is ever the horizon]). If the chase is long enough (*reasons on the march!*), who will notice that the system closes back upon itself (like a dictionary, like the hermeneutic circle), that the rules themselves are no more or less artifactual than the things they seem to govern, no more or less arbitrary than the couch, or what we do on it, that they are not a standard in the sky to which we can turn in a moment of doubt, but a voice we will into being.

Who wills into being?

All of us, collectively. The voice that is heard in the rules is that of our interrelatedness, across time (it is the voice of our class system, of our structure of differences) and through it (it is the voice of our trajectory from class to class, from generation to generation). The voice that is heard in the rules is that of our social class (or our fraction of our social class), but it is a voice whose mouth remains unseen . . . *beyond the circle of the codes.* We say it again: the circle of the codes permits the Voice of Class to speak in the rules of the home without ever being heard as such: individual behaviors seem to be required, not by the Voice that demands them, but by the codes that relay them. It is this mask of the codes that above all else naturalizes the culture of the room, that allows it to seem no more than the idiosyncratic consequence of some choices made, if not at random, then at the bazaar of history ("it was what they were selling at K-Mart that time we were there, is all"). The circle of the codes permits the room—which is the rules, turned through a certain angle—to appear as a matter of chance (nobody *means* anything by any of it).

Class just ∴ . . . *disappears* (along with the rest of the social system).

The Electric Fan

In a room closed against the heat, less hot than still, less still than stuffy, the fan's whiffling of the air is delicious.

The fan is a Panasonic AC Desk Fan, though with this on a desk, there wouldn't be room for much else. Not that at less than a square foot the base is all that big, but the fan-head sweeps through eight cubic feet, where "sweeps" is the operant word. On "HI" the sixteen-inch-diameter blade circulates a couple of thousand cubic feet of air a minute (less on "MED" and still less on "LO") and, if not impossible when oscillating, with it pointed right at you this is too much to read a newspaper by. Speeds are selected by "Fingertip Air Control" switches on the base; oscillation status by a knob on the motor housing. The sixty-cycle motor draws just less than an amp at 120 volts. The power cord—stashed in the base—is fingered in, pulled out as needed, and held in place by genuine friction. Blade and housing are plastic, the guard aluminum, the whole kit and caboodle made in Japan by the Matsushita Electric Industrial Company, Limited.

We haven't counted the number of times the blade revolves a minute, but you can tell just by looking at it that you couldn't whip the plaited fan back and forth at even the "LO" speed. A luxury, consequently, when the room's closed up against the heat . . .

rules   205. "Don't play around with it" (Chandler, Randall, ktPRO).
206. "Don't stick fingers in it or anything like that" (Chandler, Randall, kPRO).
207. "Don't cut the fan off and on, off and on" (Denis, kAPP, kCON).

With a palm frond fan (59), who needs an electric one?
The Woods.
Because, frankly, the hand fan can't cut it. Moving to Raleigh from New England, Denis and Ingrid brought with them a single window fan. But on the first day in their new apartment they looked for another (the heat was oppressive [it was like a Middle American military regime]). They finally found a General Electric hassock fan (just like Denis's grandfather's) that for years spent its summers in the living room. As their means expanded, so did their inventory of fans. The year after they moved into the

house they bought two more window fans. Then one summer Ingrid bought the Panasonic for Denis's study (though after they installed a ceiling fan there, the Panasonic moved out to the living room). They put in an attic fan. At a local industrial supply house they found a Dayton shop fan that practically blew the couch over (it's now in the basement). Ceiling fans spread like mold (only the living room lacks one now). This year they added a high velocity Patton window/floor fan (for the living room [the Panasonic's upstairs in the kids' room]).

This is a kind of hysteria (it is fan mania). But is it more than a theatrical reaction of Northerners exiled to the South? In fact, it is a measure (moreover a precise one) of the relative indifference of the house to the sultry, sticky climate in which it was built. Through the Civil War local builders raised houses suited to the insufferable summers: they had big rooms with huge windows and high ceilings and steep-pitched roofs that got the heat up and out. They were raised up and off the humid earth in shady groves and often sited to take advantage of prevailing winds; and they wore porches as capacious as sun bonnets that were almost second homes for the summer months. Those who erected Denis and Ingrid's house had something other than summer on the mind (the delirium of fashion [moreover one dictated by the North]). Even as originally constructed, it represents the erosion of each of these desiderata except for the steeply pitched roof (and here only because its particular style dictates a gambrel roof). The house is two rooms deep with skimpy porches and no hall of any kind (except, on the second floor, *surrounded by rooms*). The air admitted by the comparatively smaller windows . . . *sits*. Making things worse was a lowering of the ceilings (to save on winter fuel?). Brutally exposed to the sun on the south (where a narrow lot combines with overhead wires to produce . . . no trees); and, after the loss to the city of a monumental oak, naked to the long afternoon sun in the west as well, the house in the summer is warm at best, often hot, and occasionally infernal. During Chandler's first summer, the temperature in the dining room stayed above 90° night and day for nineteen days. Upstairs . . . it was worse.

How is this problem solved?

By plugging a Japanese appliance into West Virginia coal.

Instantly the fan as marvel of handcraft (59) is replaced by the fan as danger to the hand. This is the threat that justifies the protection coding of *don't play around with it* and the frequently frenzied *don't stick fingers in it or anything like that*. What permits this capitulation to a technology that is simultaneously exploitative and dangerous?

Though of all the ruptures between the family and the competence that constructed the house this is the most protracted and least forgiving, it is paradoxically the one that can most readily be made to disappear: the heat will be seen as *Southern*, not as an idiosyncrasy of the house (it's a *regional* detour [it takes you off the road]). This begs the question how people lived here before electricity, but by the time it's asked, the fan is on and everything is . . . *cool* (let us not close our eye to the possibility that this was a reality foreseen by the competence responsible for the house).

The fan is thus the tragic consequence of projecting "a bourgeois identity crisis" in built form (this is Henry Glassie's phrase for the succession of revivalist "styles" that obliterated folk housing in Raleigh [postmodernism is its most recent manifestation]): long after the crisis is resolved (and even where it's not), *the buildings remain* to torture and torment (final twist: the inferno the Woods occupy is praised by visitors for its *huge* rooms, *high* ceilings, *lavish* fenestration). That is, the fan represents the tragic consequences—sooner or later experienced by every kid who sticks his fingers through the grille—of evaluating the social context (in this case: the display of bourgeois status) as more significant than the physical context (in this case: the climate). Inevitably, this is utopian.

But is not the Woods' refusal of air conditioning a similarly utopian evaluation of the social (communal intercourse, human solidarity) over the physical (the stubborn fact of a sticky stuffy living room)? Indeed it is. The fan forces us to see that between the "room" and its inhabitants nothing less is at stake than a contest between competing utopias.

For the room, this is . . . *serious business.*

XXXVI · WHO IS SPEAKING?

We know that it is the Voice of Class that is heard in the rules but . . . *who is speaking?*

We are. We are speaking.

Notably, we are not saying anything we might. For instance, we are not saying, "Chip the glasses and crack the plates! That's what Bilbo Baggins hates!" We are saying, *be VERY careful* (RULE 103), *don't scratch up the floor* (RULE 44), and *don't do anything that will break the glass* (RULE 29), we are saying with Bilbo, "please be careful" and "please, don't trouble! I can manage."

How do these words end up in our mouths?

We have insisted that they adhere to the "room" and the things in it until stripped from them by the presence of barbarians. This is the same as to say that these words are implicit in the things (that they *are* the things in rule space); which is to say that they came into being with the things themselves, that they are nothing but another way of *seeing,* of *saying* the form of the thing. It is in this sense that the descriptions with which we have opened the treatment of each object are no more than translations of the rules that follow them. For instance, "The box itself, not quite eleven by fourteen inches and two deep, is carefully constructed" and so forth and so on (31) is just another way of saying, "Don't touch—just look at it" (RULE 145). The deep structure of which each is an expression is a commitment to *care* (as in "carefully constructed," as in the implication "be careful" in "don't touch").

This is not the only translation: the description after all is never more than another way of saying the name ("Peter's box"), of seeing the image, of understanding the site that it occupies in the multidimensional web of associations that is *itself* just another way of saying, of seeing, of understanding the room (the box was made by a family member, it makes connections to Mexico, it participates in the thematics of proletarian revolt, et cetera, et cetera). This name, this image, this significance, this particular genesis (which is the essential burden taken up by the description) is but the name of, the image of, the genesis of, the meaning of the rule. To *say* "Peter's box" is to say "Don't touch—just look at it." To *see* Peter's box is to hear "Don't touch—just look at it." To *understand* its position in the web of associations that converges on Peter's box is to understand "Don't touch—just look at it." To *trace* the origins of Peter's box is to unfold those of "Don't touch—just look at it."

This is of course where the words come from (the question how these *words* end up in our mouths is just another way of asking how these *things* come into being in this "room"): the words are verbal transcriptions of the intentional acts that brought the things into being and into their current coordination; they are a kind of deferred imitation (as Jean Piaget might have put it) of the motoric activity that we disaggregate into *making, getting, putting* (Peter making the box, the Woods putting it on the wall), everything conspiring to say—the paint being brushed on, the postcard being glued into position, Denis and Ingrid cocking their heads to make sure it's straight, cleaning, dusting it through the years—"Don't touch—just look at it."

The rules are not privileged (they were just our starting point). They are only *one* version (moreover one requiring the presence of barbarians to become manifest). Though the number of forms taken by an object may be limited—*name, image, genesis, praxis, meaning* may exhaust them—the numbers of translations within any of these forms are without end. Every time a rule is stripped from an object and run through a mouth it is likely to come out . . . *different.* There is no essence: only this unending play.

Is there any point to the play? Not really, but it does have another name: it is the social life in whose structure we detect the class whose Voice is heard in the rules. Does this mean that the rules are not deferred imitations of the motoric activity that we disaggregate into making, getting, putting? Not at all. It no more than insists that in these too the Voice of Class is sounded.

The Mantelshelf

Floating glossy before a lusterless wall, even from the door the mantel seems to score—in highlighting marker—an underline across the room: *note this!*

The enamel covers a rectangular frame that obscures what is doubtless a crude mantel proper—that is, the lintel that actually supports the masonry above the fireplace. Everything that we can see projects from the outer frame of the mantelpiece ten inches, as a shelf five inches deep, five feet long.

It's not an underline, but it makes a point, whatever.

rules   208. "Don't touch the woodwork with your dirty finger" (Denis, TAPP).
209. "Don't hang off the mantel" (Denis, TPRO).

The separation of fireplace (57) from mantel is not adventitious. It reflects a division fundamental to the room, that between fire and hearth, between the wild and the place that constrains it, between Nature and Culture.

This is reflected in the rules as the distinction between the dirty and the clean (an opposition fundamental to the floor [8]). If the kids note *we shouldn't play in the fireplace* (RULE 202), this is because it'll get them dirty; if they are enjoined from pulling the chain (RULE 203), this is because the damper will shed dirt (it'll get them dirty [it'll get dirt everywhere]). No matter how carefully cleaned, the fireplace remains a scorched site—that is, one of ash, cinder, soot, smudge (that is, marked by Nature . . . *at its worst*). In opposition, but immediately circumjacent, a site of the clean: *don't touch the woodwork with your dirty fingers,* an injunction recognized as the inevitable companion of any sign of the Patriarch (door [2], doorframe [3], windows [4, 5, 7], newel post [10]). Since the dirty and the clean abut here, the threshold of significance is unbearably high (that is, marked by Culture . . . *at its finest*). Unavoidably, the prestige of the dancing god is absorbed by the mantelpiece. It is reradiated as *gloss* (etymology: flame, spark, the glowing).

In this way, Culture subsumes . . . the Natural.

The Piece of Wire on the Mantelshelf

Volute of copper, curpous curl, brazen djinn . . .

Seven ounces of copper wire, forty-two inches long, clipped at both ends and bent into this twist of a coil . . .

Spiral of cuprum, cupric whorl . . .

rule   210. "Don't play around with anything on the mantel because you could break some of them" (Randall, kcon, tpro).

The wire on the mantel is the touchstone in the room of the "find," of that "*trouvaille*" which José Pierre describes as the reply to a generally unformulated question.

Ingrid found the wire at the bottom of the hill behind the Irregardless Cafe. When asked (recently) why she picked it up, she said, "Because I knew it would be beautiful . . . Because I knew you'd want it . . . Because it was copper." When asked why that was important she added, "Copper that thick . . . is . . . I don't know what. Beautiful . . . What if you needed it?"

Denis stuck it on the mantel, clearing off who remembers what (though whatever they were, they formed the object of *don't play around with anything on the mantel because you could break some of them* [spoken in Randall's finest collaborator's voice]). The piece of wire has *not* been there consistently since. It comes and goes like Webern's music, whenever there is a need for clarity.

As a "*trouvaille*" it makes connections in the room to the Ernst (20), Peter's box (31), and Denis's collages (32, 33). As an assault on the bourgeois world implicit in the mantel—it is a piece of junk on the hearth of honor—it is linked to everything in the room that rises to this contestation (6, 20, 21, 24, 25, 29, 31, 32, 33, 37, 39, 40, 41, 53, 55, 56, 59, 60, 61, 64, 65, 68). As the most self-conscious—yet innocent—of all these gestures, it may be taken as their touchstone too.

Ironically—after all!—on the mantel . . . an object of significance.

Peter's *Genovevo de la O*

On entering the room, a sudden change in the weather: over the mantel a dazzling rainbow spreads itself. That it subsists in mountains only makes it the more fabulous . . .

It is Peter's portrait of Genovevo de la O (fabulous name!), made in 1970 with oils on board, using nothing but his fingers to paint with. At better than thirty inches by forty, it fits the wall it's on as though painted for it. It was framed by the Bonfoey Company in Cleveland, who applied the gold leaf by hand.

Even the colors of this rainbow are rocky, *ultramarine* blue, *cadmium* yellow. How wonderful their transformation . . . *into light.*

hand-off   2. "Same as every painting" (Chandler, Randall, to RULES 90, 91, 145).

We know in talking about his *Malatesta* (41) that Peter claimed for it a religious quality he denied the *Genovevo de la O* and most of his others, regarding them frankly as works of propaganda in which, aside from any other value, his interest lay in getting people to ask about their subjects and so be led to "read and think about changing the world." But just as we saw in the *Malatesta* a work of propaganda, so here we find a religious painting. Peter himself is explicit about the halo: "The hat is a halo, that's clear out front. Hats and halos were connected in my mind." Reference to the source photograph in John Womack's *Zapata and the Mexican Revolution* demonstrates the extent to which Peter was obliged to distort de la O's sombrero in order to free the halo (to say nothing of his reworking of the fret in the braid to the release the rays of the nimbus). Peter had to reach back to the later Middle Ages to find a model for the halo as a flat disc (though the Christ's head in Stefan Lochner's *Virgin in the Rose-Bower* is wonderfully close) and even earlier for the direct gaze. He was more likely, however, to have drawn on folkloristic sources for his treatment of the head, as he certainly did for the mountains and the church. There was no background in the Womack photo, so Peter had to invent one: "You know I'd been selling a lot of folk art at one point, and I got it from that. The mountains are straight out of . . . I don't know what. They're probably just fake folk art. The church *is* fake folk art," that is, painted at a level of sophistication less than his own, in the manner, for example, of the tin *retablos* to be found in churches throughout Mexico.

Again, then, this is a reredos, an altarpiece (given its location above the altar of the house, inescapably so), like Peter's box (31), though probably intended less to instruct or arouse than to exhort to virtue (by example: the project is Plutarchan). In contradistinction to the triune halo of God the Father or the cruciform halo of Christ, the circular halo belongs particularly to Mary, angels, and saints. And certainly, were it possible to call any guerilla general a saint, Genovevo de la O is the one. A leader from the village of Santa María (appropriate origin for a man with a halo), de la O was among the first to "take to the hills" (dramatically represented in the painting [they are like the attribute of a saint, Jerome and his lion, de la O and his hills]). Womack notes that "the bands were small and poorly armed: the group Genovevo de la O collected in the mountains north of Cuernavaca had only twenty-five men, and only de la O had a gun, a .70 caliber musket." Yet de la O was to become the most effective Zapatista general, and the one most serious about agrarian reform: "Land and Liberty, and Death to the Hacendados!" (like any Mexican revolution, this was essentially against the hereditary nobility's monopoly of the land). Unlike many guerilla leaders, de la O never lost his sense of who he was: "When Zapata entered Cuernavaca on May 26 he intentionally came in not where Asúnsulo had gone to receive him but where he knew de la O was waiting. And there he found him, a gruff, stumpy fellow, dressed in a farmer's white work clothes, lost among his men; Zapata was deeply impressed." And so was Peter (who has made of the white work clothes the most brilliant thing in the painting).

Carrying into the field of military action much that Malatesta was calling for (at the same time: Zapata's Plan of Ayala is announced the year before Malatesta is tried in London), de la O infuses the room with . . . *a call to arms.* In his hills (attribute of the outlaw), with a machete at his back (detail wholly original with Peter), his gun in hand and bandolier around his neck, he is the very emblem of political *action* (the land is yours, Zapata told the Indians, and the only way you can get it is to take it). Praxis over theory. In the room, Malatesta and de la O complement each other, together whole, the compleat political actor.

How much do the kids see? Every day more (just last week Chandler asked why the hat was so funny). By the time they are ready to pass on the culture they are learning (childhood is an apprenticeship) they may even appreciate the implications of the *rainbow* throne of Christ at the Last Judgment.

The rich man will not make it; the outlaw might.

XXXVII · PROPERTY

The room is a great debate. On one side: for property. On the other: against.

In general the "room" is on the side of property. Taken as a whole it is the most public sign of a certain propriety that takes property for granted. Through its endless excesses it declares those means that presuppose the possession of the land on which the house is built, as well as the surrounding land that impregnates *that* land with its full meaning (the house of the Woods sits back on its lot [it would like to pretend to a circular drive {but it doesn't have enough land (the entire show—as we know—is a sham)}]). This lot is not really the property at stake in the debate, although its exclusive possession by the Woods is at the very least *symbolic* of that being debated, and it in fact encapsulates the essential features of the ownership around which the debate will swirl.

The "room" announces its position in the syntagm of the door (2, 3, 4, 5) whose capaciousness prepares any entrant for the volume of the "room," which is immediately understood as being . . . *too big.* What is the space for? (We know it is not for cooling.) *It can only be to display its excessiveness.* This instantaneous comprehension (it has something in it of cynicism) avoids the agony of confronting the fact concretely: what the size of the room cannot hide—what it is at pains to insist upon—is its undue share of resources (floorboards, joists, beams, planks, nails). What justifies this excess in a world where others suffer deprivation? Modesty forbids, but candor insists: *only innate superiority* (those inhabiting the "room" *deserve* more [because they are more *deserving*]). This excessiveness, justified by an apodictic superiority, pervades every sign the "room" deploys (each is too much): the contraption of the door (2), the lace of the doorframe (3), the crystal of the window (4), the diamonds of the sidelights (5), the glass of the floor (8), the bristling of the newel post (10), the ambivalence of the landing (14), the privatizing of the stair (15), the brooch of the little window (17), the superfluity of the couch (30), the luxury of the windows (47, 50), the fraud of the fireplace (57), the gloss of the mantelshelf (62) . . .

Nothing in the room escapes being contaminated by the pollution of this insidious utterance. Thus the coffee table ensemble (30, 35, 53, 65)—to pounce on an egregious example—presupposes the kind of leisure that bespeaks the means implying a certain level of property (it is the dying echo of the withdrawing room to which seventeenth-century women retreated after dinner so that the men could drink, smoke, and swear at ease). The splendid whiteness of the couch points toward a kind of manners (no dirty hands [no auto mechanics in residence] but if so, then the means to clean them) as the absence of a television gives this luxury the form of a particular class (its members engage in polite conversation). Tainted by the contagion of the glass and glossy, records, even paintings of anarchists and revolutionaries, are seen to allude less to a pleasure or polemic than to a certain level of education (appreciation of the arts [no pressure to drop out of high school, no competing demands on the resources needed to attend college]). In such an atmosphere, folk "art" is no more than a trophy of a voyeuristic and consumerist travel. Infected by the virulence of the syntagm of the door, even the unused lock and uncurtained windows appear less the signs of an openness and trust than the arrogance of a sense of security that in a world of unequally distributed resources can only be spelled *p-o-l-i-c-e.*

These, of course, are actually present in the room in Peter's *Malatesta* (41), a painting of a man who wrote:

> The basic function of government everywhere in all times, whatever title it adopts and whatever its origins and organization may be, is always that of oppressing and exploiting the masses, of defending the oppressors and the exploiters; and its principle, characteristic and indispensable instruments are the police agent and the tax-collector, the soldier and the gaoler.

The room is hereby rendered *self-conscious* about the property reveled in by the "room," a self-consciousness that suddenly is evident in everything. No case need be made for the call-to-arms over the mantel (64), for the egalitarian transcendentalism of the Strand (56) or for the memento mori of Peter's box (31), but in their light the revolutionary claims of surrealism in the Ernst (20), Denis's collages (32, 33), and Yunkers (40); of the naked metal in the lamps (34, 42), chairs (29, 53), and table (65); of native Americans in the Cherokee basket (24) and Lacandon drum (39); of the child in the toys (25, 60) and kids' books (28); and of the worthless in the wire on the mantelshelf (63) acquire fresh presence. This suffuses the white walls (15) and plain floors (8) with that Esprit Nouveau of the essential and necessary which tints even the white of staircase and doorframe. *Nothing* escapes. Among the most potent of paradigms saturated in the room, *old/new* even recuperates the "room" for the revolution as a historical (that is, transcended) stage of development; and from this perspective even the bourgeoisie "requiring" this room (Denis's grandparents) seem no less victims than those whom its excesses deprive of resources for their own living.

What stand does the room take in this debate?

None: it *is* the debate, *both voices,* at once the bells on the door (6) mocking the pretensions of the door (2, 3, 4, 5) *and* the door mocking the pretensions of the paintings and the records, the stainless steel furniture and the Mexican toys. Where then do the Woods stand? Evidently, since they live the room, they live the debate. Alternately smug and guilt-ridden, wanting this but convinced they already can do without that, gnawed by financial worries but certain that they earn too much, contemplating at one moment the plight of the really poor, at the next the pornography of the obscenely rich, they flop from position to position, liberal, indecisive, anxious, confused.

The room they live? In spite of everything . . . *middle class.*

Mies van der Rohe's Coffee Table

Magic moment: to come from the kitchen in a dark house and find moonlight on the coffee table . . .

This is Ludvig Mies van der Rohe's Tugendhat coffee table (first called Dessau, lately Barcelona). It is a slab of glass a thousand millimeters on a side (these numbers are not so round in inches) and twenty thick, sitting on a frame of welded steel bar thirty-five millimeters wide, eleven deep. This has been worked by hand, and polished to a mirror finish. The frame forms a cross—nine hundred and fifty millimeters along each beam—that lofts the glass five hundred and thirty millimeters into the air. The whole thing is strong enough to support Denis (who does a jig on the table to prove it).

Into the beam over each leg a hole has been drilled: the glass sits on the four white plastic studs that plug them. A small transparent disc slips over one of these to

65

compensate for a floor anything but flat. Twelve bevels take the edge off the edges of the glass: it's sort of an emerald cut, but with an *enormous* table, long thin bezels, four girdle facets eleven millimeters deep, long thin pavilions, and a culet the size of the whole underside of the table.

Whatever the cut, when it's moonlighted late at night, the table is that form of feldspar we instinctively call . . . *moonstone.*

rules 211. "Try to keep your fingers off the table" (Denis, ktAPP).
212. "Don't play with anything that will scratch it up" (Chandler, ktPRO).
213. "Don't put your feet on the table" (Chandler, kAPP, kCON).
214. "Don't leave your stuff on the table" (Denis, ktAPP).
215. "Don't leave *Tintins* on the table" (Randall, kAPP).
216. "Don't lift glass part off" (Randall, tPRO).

At the heart of the room: an emerald-cut diamond in a setting of platinum.

This is the meaning of the glass, the bevels (it is a gemstone), the mirror finish on the stainless steel (it is the prongs, the collet of the setting), that it is a diamond ring (the room has lace at its throat, a diamond brooch, a graduated *rivière* . . . a diamond ring on its finger). Coffee table, sign of an effete transformation of the drawing room, will take the form of women's jewelry (the living room is a Georgian drawing room). Yet the pieces will not constitute a parure. Though the lace collar, brooch, and *rivière* may go together, the ring does not. Seemingly caught up in the syntagm of classical patriarchal Western culture (Georgian revival door, Georgian drawing room), the table at the same time manages . . . *to twist away* (despite its recuperation as lobby icon), elude its grasp, regain its feet as a mark of revolutionary purity. The oppositions are familiar: the revealed structure invites the Expulsion from the Garden (and Hitler *will* force the modernists from the Reich); the exposed metal mounts a telluric assault on the uranian ascendency (metals are ever the mark of the mole, the dwarf, the *chthonian* [as the eagle is that of the sky {the Reich}]); though substantial, its appearance is ethereal (this is its elegance), but elegance is always suspect in the "Blut and Boden" world of the burgher [only the solid can be counted {on}]). In the cozy warmth of a family parlor, where is the room for the icy purity of this embodiment of a diagram? Has anything ever distinguished with more sovereign authority the support*ing* and support*ed*? This is not the lounge chair (53): nothing is hidden.

Well, almost nothing.

The machined impression responsible for the aura of the mass-produced (the industrial [the modern]) is not entirely justified; welding, finishing, and polishing are done by hand; the table remains *very much* the exquisite product of a handicraft tradition. (Even the *metal/ wood* opposition can be dissolved (or relocated): the table was originally offered in a choice of glass and . . . *rosewood.*) Table for a bourgeoisie with a taste for the cutting edge? Mass-produced machine for a new way of life? Wonderful mystery: that anything so clear and simple can be so hard to pin down (ineluctable excess of the signifier).

What will the table be for the kids? A precious object certainly (*try to keep your fingers off the table, don't play with anything that will scratch it up*). A site of proprieties (*don't put your feet on the table* [everyone does, but the table will project some intimation of the room's Georgian origins]). Also, one of a precious candor (*don't lift the glass part off* [more than anything else in the room, the table . . . *shows all*]). But most of all it will be the site of a rare distinction: no television (the local newspaper will run a story on families without television: the photo will show Denis, Randall, and Chandler on the couch [30] staring at a book propped open on the coffee table).

The table will embody this absence (handily). It will be experienced in the room as a dual freedom from mystification. It will make of the room the proclamation . . . *of an independence.*

XXXVIII · PASTORAL

The room is a great concord.

This is the consequence of the prominence in the room of the evasive tone of the saturated paradigm. If the room can imagine a distinction, it will embody the values on both sides. In this way the distinction is not denied: it is absorbed. A potential stridency is turned . . . into a part song, into a madrigal.

For instance, if there is a distinction between *painting* and *photography,* the room will exhibit paintings (41, 64) *and* photographs (56). Whenever possible the room will expand the terms of a distinction. Here the paradigm is enriched by the introduction of the glissade terms *drawing* and *print,* diminishing by steps the sense of distance separating the "nobility" of painting from the "commonness" of photography: *painting* (41, 64)/*drawing* (37)/*print* (20, 40)/*photograph* (56). At the same time, linked through the lubricity of "common," two further paradigms are saturated: *noble*/*common* (at stake here is the question of skill and the ease of its attainment) and *rare*/*common* (here it is a question of edition size). These could take us to *easy*/*hard* and so on around the hermeneutic circle.

There is no privileged point of entry. We might have dived as readily through the moon pools of *figurative* (37, 56)/*abstract* (40), or *realistic* (41, 56)/*imaginative* (20, 41), where glissade terms are no less available (for instance, the Ernst [20] *is* figurative, but scarcely; while the Rozzelle [37] photorealistically depicts an encounter from a

dream). Because so many paradigms in the room are saturated it is possible to get wherever you want from wherever you happen to be. From *painting/photography* to *glass/mud*? No problem. We already know how to get from *painting/photography* to *noble-rare/common* to *hard/easy*. *Hard/easy* is a direct translation of *high order-energy/low order-energy*, which is one way of conceptualizing the difference between *glass* and *terra cotta* (*high-fired/low-fired*). Less fired than terra cotta is *earth*. Still less fired is *mud* (water is what the heat drives out). The paradigm then runs *glass/terra cotta/earth-dirt/mud*. But are all the terms embodied in the room? We know the floor (8) as a cooked-earth is stridently anti-mud (*don't walk on it with muddy feet* [RULE 51]), but the room nevertheless *makes its own* every time Ingrid waters the plants (it does this with the regularity of the monsoon). A saturated version of one possible expansion of the paradigm might thus run *floor* (8)/*flower pot* (43)/*potting soil* (43)/*mud* (43). Note, by the way, how the route from *painting/photography* to *glass/mud* passed through *noble/common*—that is, through *propertied/not propertied*.

Does the room introduce terms to paradigms which it fails to saturate? *Black/white, wood/metal, handmade/machine made, primitive/sophisticated, primitive/folk/elite, folk/popular/academic, sensuous/intellectual, aboriginal/non-native, of regional origin/of national origin, American made/foreign made, old/new, revanchist/avant garde, conservative/liberal, left/right, Nature/Culture*—the room saturates them all, and if in varying degree nonetheless completely (even the mining of nature by the houseplants seems less empty when all five are seen together [there is just the slightest suggestion of . . . *a jungle*]). Though equally evasive of cynicism and irony, the evasion accomplished by obliteration is not that achieved by saturation. Obliteration *denies* differences (it understands *let them eat cake*). Saturation *acknowledges* them (for example, that there are different economies) but denies that genuine values (those cherished by Denis and Ingrid) are homologously distributed in everything they touch (Lacandon forest dweller, Oaxaqueño peasant, North Carolina factory worker, world-wandering waiter, all equally produce beautiful artifacts: drum [39], car [60], chair [29], painting [64]). Because of this *comprehensive* acknowledgment the paradigm is fatally compromised (it is occluded), but without denial or being forced into cynicism or irony. It is a purely intensive tone: *no, really!*

It is precisely this tone that Kenneth Burke identifies in William Empson's conception of the pastoral, a literary genre that Empson insisted was "felt to imply a beautiful relation between rich and poor." Burke comments:

> True, whereas the "proletarian" critic's emphasis upon "class consciousness" would bring out the elements of class *conflict* [as in our great debate], Empson is concerned with a kind of expression which, while thoroughly conscious of class

differences, aims rather at a stylistic transcending of conflict. We might say that he examines typical social-stylistic devices whereby spokesmen for different classes aim at an over-all dialectic designed to see beyond the limitations of status.

And is this not what the room does, attempt to transcend the conflict (to end the great debate by evading it), attempt to see beyond the limitations of status (instead of getting trapped by them)?

The room: *pastoral.*

Chandler's Turtle on the Table

Lumbering from behind the terra cotta of the lily's pot (67), heading across the table toward the couch, a reptile of the order *Testudinata.* Genus? *Terrapene.* Species? *Ludus litterarum "papier-mâché."*

This is a school project of Chandler's, a project of Mrs. Hood's model-making class. Beneath the poster-color paint is a layer of papier-mâché that unites in the form of a tortoise two foil pie tins, five spools (of thread), and the bottom of a single well from an egg carton. An inconsequential bagatelle whose purpose in being was to be made, it nonetheless displays—like everything else in the room—the sophistication of an advanced industrial economy (flour, newsprint, aluminum, wood, plastic, gum binders, pigments). That its existence threatens that of the entire order *Testudinata* is an irony Mrs. Hood doubtless left . . . for someone else's class.

What is this doing on the table?

Resting.

It's on its way from Wiley Elementary—where Chandler made it—to the closet upstairs in Denis and Ingrid's bedroom. Chandler says: "I brought it home. You wanted it. I always come home and say, 'Do you want this?' You did."

Why do Denis and Ingrid say *yes?*

It is a gesture of respect. It is what they expect the kids to show all the things that are the room in its artifactual presence. It is what they want the kids to feel for the *making* (it is what they feel for Chandler's making). In fact, they have little other interest in the turtle (this is not inevitable: Denis has had a construction of Chandler's on a wall in his study for years; and airplanes made by Randall are an integral part of the middle room on the second floor). Consequently, it will not be here long.

The room: way station, shop window, gallery for traveling exhibitions.

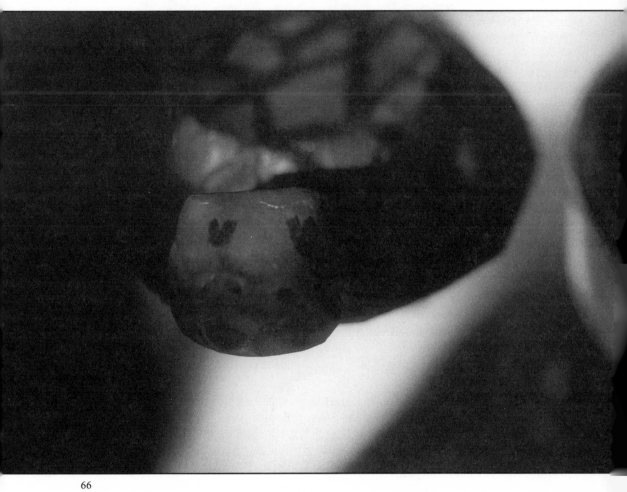

The Plant on the Table

The mirror of the table turns into fifty-eight the twenty-nine leaves of the lily. Out of the plant, it makes a jungle.

This is a daughter of the peace lily by the speakers (43), maybe twenty-six inches from side to side, twenty-two or so from top to bottom. It's in a terracotta pot six inches across in a terracotta saucer six inches across in a plastic liner seven inches across. This last is from the Curtis Wagner Company of Houston, Texas.

When the fan passes it's like a hurricane in the tropics: leaves every which way . . .

hand-off   7. "Same as the rest of the plants" (Chandler, Randall, to RULES 84, 85, 86, 99, 181, 182, 183, 184).

Another peace lily (they can take the Santa Anas of the fans).

What is it doing on the table?

Standing in for the cut flowers the Woods cannot afford (tulips lavish in a crystal vase of Alvar Aalto's). Cut flowers would conclude the theme of the plant: an *exclusively* decorative Nature (*dead* is the zero degree of the *tamed*). But because dead flowers wilt, they must be endlessly replaced: this makes of them the perfect site for a bourgeois luxury.

It is this regular replacement the Woods cannot afford. Instead, they water their lily. *Noblesse oblige!*

XXXIX · A ROOM LIKE ANY OTHER

In what sense is the room like any other?

In that it is brought into being out of choice. It is in this sense that it is impossible to distinguish it from *any* other room.

But in talking about this with people, we have found many who cannot agree. They refuse to accept that the environment, especially that most proximate to them, flows out of choices that they make. To be sure, these are not unconstrained, but to hear them tell it *everything* "just happened by chance" or "it was the only one they had" or "my mother gave it to me: what was I supposed to do, throw it away?" What we find especially remarkable is the way this fatedness erases every other quality: color, form, size, weight, use, *all* are absorbed by this inescapable history that elects the thing to be a part of their room, beyond choice or personal responsibility. Talk, even about things in *our* room, inevitably circles back to this singularity: *but your brother gave it to you,*

didn't he? And once this is granted, there is nothing else to say . . .

The *fact* is that Denis's brother has given Denis and Ingrid many things. Some of these are no longer even in the house. Among those that are, many remain in the upstairs closet that is soon to receive Chandler's turtle (66); and among those that aren't, only four are even on the ground floor (a peculiar chain of handmade paper hangs in Denis's study). Among the three in the living room (31, 41, 64), only *one* hangs above the mantel (64). Does this mean that it is irrelevant that Peter gave the painting to Denis and Ingrid? Not at all: we deny claims of neither history nor family (besides, Ingrid explicitly referred to the *way* Peter gave it to them—"you know, that way Peter has"—in talking about what she would save in case of fire). That Peter gave them the painting *is* essential, *but it is not sufficient.*

What is?

In fact, nothing is. Or at least nothing less than *everything.* Certainly Genovevo de la O is implicated (it's his portrait after all), and he drags along the whole Mexican revolution, which necessarily implies the land tenure system implaced by the Conquest, which without even trying lands us in the New Spain of the fifteenth century (with the starlight through the window, the room tangles with eternity). But isn't this our point, that the entire history of the world is necessarily implicated in the presence of Genovevo de la O over the mantelpiece in the living room of Denis and Ingrid Wood? (That nothing in the room is less than a point of departure into that history?) To say this is not, as it might appear, to exempt Denis and Ingrid from one iota of responsibility for the painting's presence there, but to insist that this entire history is brought forward out of the past in the lives of those living it in the present. Global economies, social class, each is made manifest in the world through those alive in it, who through their responsibility and meaning-making bring it into being. The world put Genovevo de la O over the mantel, but only because Denis and Ingrid did.

And every room is like this.

It is true that not every room has a painting by Peter in it, or his grandfather's easel, or a photograph on the wall by a man whose photographs were exhibited by the mentor of the man who gave Peter's father a place in *The Family of Man,* but every room has a wall with something on it, or with nothing on it, and in both cases, this is the way the history of the world is being lived in the present by those who live in the room with that wall. It is rough timber or thatch of straw, it is plastered and papered or taped and painted, it is stuccoed or mosaicked. We are not oblivious to the reality that constraints vary (and systematically) both in force of compulsion and in the forms they permit or encourage, but no room can escape being the resultant of a living unfolding in space and time.

A room like any other—"But it's not," Chandler says. "It doesn't even have a carpet, *dummies.*" And *this* is how it lives its history: no carpet, but smart-aleck kids. *Exactly!*

XL · PRIM IRONY

As self-conscious as it (all too) evidently is, what tone does the room take toward the history that it lives? It is one of "prim irony." This is a salient of William Empson's pastoral that Kenneth Burke describes as "an attitude characterizing a member of a privileged class who somewhat questions the state of affairs whereby he enjoys his privileges; but after all, he does enjoy them, and so in the last analysis he resigns himself to the dubious conditions, in a state of ironic complexity that is apologetic, but not abnegatory." This is the stance of the room (it is that of the authors of this book).

In spite of everything . . . *ironic detachment.*

The Baseball Game

It's two to one with one out in the fourth inning in Mies van der Rohe Park here on the Barcelona table. There's a batter at the plate—the home team's up—and another in the on-deck batter's circle. The pitcher's winding up, the catcher's in his crouch . . .

The nineteen wire-and-pottery figures, scoreboard, and bench came from Guatemala via Canada where Denis brought them in a little shop in Vancouver's Point Grey neighborhood where he happened to be staying with his friend Tom Koch. They were in the window with a similar bunch of soccer players. None of the crudely painted figures is more than two and a half inches high or weighs more than a quarter of an ounce. The whole thing has the air of a folk art cooperative organized by some well intentioned *norte americanos.* Certainly it is not the work of a named folk artist, but beyond that its very inspiration seems somehow . . . *manufactured.*

Which is fitting. So's the breeze from the Panasonic fan taking off a little of the heat beating through the window with the late afternoon sun.

rule   217. "We probably shouldn't play with it, but we do" (Randall, ktCON).

Nothing in the room is less than a point of departure into the history of the world implicated by its presence. "Baseball," begins the *Britannica,* "belongs to the extensive

genus of civilized athletic competitions stemming from the primitive play urge to hit a fragment of rock with a club."

Is this what the baseball game is doing on the table (65), recalling on its sophisticated modern surface the primitiveness of our ancient face? No, like the turtle (66) it's merely passing through, showcased for the moment as the trophy it doubtless is before slipping into storage. It is a souvenir (etymology: to bring to mind) without being a souvenir (that is, it is intended to bring to mind more than the occasion of its acquisition), but nonetheless, not meant for the living room (or rarely). Therefore it is more closely connected to the turtle (both are just passing through) than to the Ferris wheel (25) and wooden car (60), with which it shares an identity as toy; or Ferris wheel, wooden car, and king (55), with which it shares an identity as folk craft from south of the border; or Ferris wheel, car, king, and Cherokee basket (24), with which it shares an identity as folk craft. As its subsequent history would demonstrate, its real affinities were with the *miniature* (at the moment it is completing a ten-month strand with other miniatures on the knickknack shelf just inside the arch to the dining room).

What we see here is the table as the *traveling exhibitions gallery* of the living room (it is where gifts are unwrapped), which thereby emphasizes the room as museum (etymology: memory).

The room *is* a memory . . . of *everything*.

The Ceiling

There are afternoons in the spring and fall when the sunlight scattered by the bevels in the table paints rainbows on the ceiling of the room . . .

These spread across a thin film of titanium dioxide and silicate particles that is secured to the ceiling in a matrix of vinyl acetate/acrylic resin which in turn adheres to a similar film brushed onto the heavy paper skin that is wrapped around the gypsum plaster and ground-up newspaper that constitutes the heart of the wallboard. Six four-by-twelve panels and five smaller pieces—suitably taped and mudded—have been nailed to a grid of joists and a heavy furring of one-by-fours on twelve-inch centers to cover the 341 square feet of plaster ceiling floating fifteen inches above it. This had had paper pasted to the finish plaster that had been troweled over a substrate of one and three-quarters tons of gypsum plaster and plasterer's sand squished between the gaps in the mesh of lathing installed to hold the plaster in place. Twelve hundred linear feet of the long narrow strips were nailed to the floor joists of the kids' bedroom on the floor above.

Of this cranky contraption, nothing is visible but the heads of six nails that have broken through paint and joint compound to star a space of ceiling as Orion's belt the sky.

Celestial phenomena . . . *where they belong.*

rules 218. "Don't touch it—we'll get it dirty" (Randall, ktAPP).
219. "Don't put hand up on it" (Randall, ktAPP).
220. "Don't throw anything at it" (Denis, ktAPP).

What is the ceiling that nothing should be thrown at it?

It is not the sky.

That is—despite its etymology—it is not "the clear hyaline, the glassy sea," not the infinite blue serene, not the cloud-flecked welkin shivering with rain, sleet, and snow, not the wind-scoured vault, the chalky oven, the phosphorescent furnace, not this cope star-studded, "this majestical roof fretted with golden fire," this firmament, this sky, this heaven. It is nothing of the exosphere, the ionosphere, the stratosphere, the troposphere. It is nothing of the atmosphere. It is not of air. It is not of nature at all.

It will constitute itself an anti-sky. Where the sky is hyalescent, the ceiling will be both matte and opaque. Where the sky is blue, the ceiling will be white. Where the sky is proteiform, the ceiling will be changeless (or only a little more grimy year after year). Where the sky is immaterial, the ceiling will be indurate. Where the sky is deep, the ceiling will be flat. Where the sky may rain, the ceiling will be always dry. The ceiling will be a matte, opaque, unchanging, indurate, dry, flat whiteness. It will be paper. It will be harder than paper. It will be chalk. It will be marble.

The ceiling will be marble.

What is marble? *It is cooked earth.* To make a marble a marl is subjected to the pressures of deformation, high temperatures, both. The room will be ceiled with cooked earth. The marble here is metaphoric, but the room *is* ceiled with titantium—mined from the earth, smelted (that is, cooked)—and silicate particles in a matrix of plastics derived from oil (of vegetal origin, but pulled from the earth) that has been . . . *cooked,* all smeared on panels of gypsum—yet another mineral—below a ceiling of plaster, that is, of *burnt* gypsum. The room aspires to marble ceilings (it would be another Greco-Roman touch [they would be white marbles from Mt. Pentelicon, from Carrara]), but it settles for a partially dehydrated calcium sulfate enlivened with flecks of titantium (this is its economy). This bears a double burden, for the ceiling is obligated to participate not only in the paradigm *ceiling/sky* (that is, *culture/nature*), but *ceiling/floor* (that is, *feet/head*). Therefore, despite the fact that both fall on the same side of the virgule in *raw/cooked,* the ceiling is nonetheless forced to constitute itself an anti-floor (the room cannot afford to obliterate the paradigm [it can't afford the marble]). Now we know that the cooked clay of the floor (8) results in *glass* which is, precisely, *hyaline,* that is, *sky.* The earth will be sky. Therefore, the sky will be earth. That is, the ceiling will be *marble,* which is to say *matte, opaque,* that is, *dull, inert,* that is, *earth.*

The mud above, the sky below: the double inversion is a double renunciation, not only of Nature but of all those who *live* with the sky above, the mud below, that is, the forest dweller (the Lacandon), the peasant (the Oaxaqueño), the bum (Peter). The ceiling is that of the gentleman with the means not only to make the world *his own,* but to make his own *world.* It reeks piety, even as it usurps it (with nature down the tubes, so is natural law [it will be reinvented to justify the situation *in statu quo*]). There is nothing new here. Civilized man will surround himself with the cooked (civilized man *is* the cooked [he is cooked in the oven of the rules]). He will sign this as the clean (the dirty plate has returned to nature). The clean is nature cooked (it is cooked in the oven of brooms and brushes, of hot water). To demonstrate its cleanliness the ceiling will have to be a surface capable of receiving the mark of the dirty . . . *to prove that it is clean.* Therefore, the ceiling will *expose* its cleanliness. Much is at stake. In the perfect ceiling clean and dry—it does not drip, it is not stained by leaks from overflowing toilets, tubs, stopped up sinks—is written a maxim of civilization: civilized man is he who knows to come in from the rain (to keep a roof over one's head: this is what one labors for [a second floor, especially one with an attic, inscribes this additional sign, that the ceiling of the living room is removed from the sky by the ceiling of the second floor, by that of the attic, by the roof, that is, is *trebly, quadruply,* a ceiling]). Any stain throws this Culture into doubt. The slightest mark is never less than the prelude to the hole admitting wind and rain—that is, weather (that is, Nature). Therefore, *don't touch—we'll get it dirty, don't put hands up on it,* and *don't throw anything at it* (moreover, *don't throw things in the living room—you might hit something* [RULE A] and . . . *what were your hands doing up there anyway?*).

As for the nail heads, they no more than attest to Denis and Ingrid's motives (who spend more money on records than on fixing the ceiling [the professor is still a student {this is student housing}]).

No, really!

XLI · THE PLAY OF THE SIGNIFIER

Every net is mostly hole. No matter which we throw over the room, effortlessly the signifiers slither free. To see the white ceiling as cooked is to miss the modernity of its silence. To see it as dropped is to overlook how it floats above the floor. To see the

Mexican toymaker in the Ferris wheel is to miss the American engineer. To see in it the World's Columbian Exposition is to miss Oaxaca. To see in the glass in the door the presence of police is to miss the flare of the diamond in the bevel. To observe in it the silica and soda, annealing ovens and polishing machines is to fail to see through it to the trees across the street.

Nor is it a question of seeing both together. For every two there will always be a third, for every three a fourth, for every four a fifth. There is no stopping the play of the signifier, which knows neither total nor essence but, as in any game, *only the next move.* If in the room the white is now on the side of the bright and the glossy, soon it will appear an ally of the essential and the necessary. We make the *meaning* as we live the room for which it is no more than another name, the meaning—unendingly transitive—ceaselessly rising from the room as through its living the past is brought forward into the present. This is not to say that the room has no meaning, or that it has any meaning, but rather that whatever the meaning the signifier *always* exceeds it, always posits more, always points out yet another route into yet another constellation of oppositions and affinities.

It is in the excess that children find *themselves,* escape the meanings imposed by their parents, become meaning makers in their own right. Involved is neither rejection nor mimickry, but through engagement of the unendingly transitive the articulation of . . . *another route.* And the stars may be the same, but in their largess they may be seen again another way. Only in the play this excess affords can kids not only grow, but make the meanings that will make the difference . . .

The room will work . . . in its play.

The Light Switches

In the shadow of Ron's drawing (37), half-hidden by the large green hands of the monstera on the cabinets just inside the door (22), is a vague umbrageousness, a couple of small darknesses, a heavy smudge, some discoloration . . .

These are the light switches, three of them behind a single paint-encrusted cover plate that measures little more than four by six inches. Six screws secure the cover to the switches, which are screwed in turn to the box that is embedded in the wall. This contains the connections that enable the paint-covered on-off levers protruding from the cover plate to route the electrical current here, or there. Cable, box, switches, cover plate—even its location—are standardized to reduce the chance for shock and fire and to enable the room to merge with the wiring of house and neighborhood, city and region as smoothly as possible.

As when entering the room in the dark, to be able to just reach over and with a flip of the switch tie the light on the porch to all that West Virginia coal, without even getting dirty . . .

rules   221. "Leave the light switches the way they are" (Denis, tcon).
222. "Don't play with the light switches" (Denis, kapp, ktcon).
223. "Don't mess with one of the switches, because it turns off the upstairs light" (Randall, kapp, tcon).

Here, three levels of codes.

On the first, a primary one in an electrical age, *up/down* is coded *on/off*.

On the second, three switches—*left/middle/right*—correspond to three different lights: *porch/upstairs hall/couch* (34). Two of these are additionally controlled by their own switches (upstairs hall, couch lamp): an ambiguity, therefore, about the significance of the positions of these switches by the door, for *on* may fail to mean *light on* but only the movement of an invisible power that we imagine flows through hidden circuits. It is to annul this potential ambiguity that the rules intervene. *Leave the light switches the way they are* allows current to flow to the switches *on* the lamp and *in* the hall upstairs, permitting them always to be controlled from their own switches. Randall's *don't mess with one of the switches, because it turns off the upstairs light* embodies the problem in the rule (he turns it into a lesson in electrical engineering).

Don't play with the light switches similarly addresses this situation, but by promiscuously embracing the porch light as well (controlled solely from this panel) it at the same time directs attention to the third level, that of the light itself. We know that the state of the door and the living room light constitute a sign of the family's willingness to receive visitors, but the porch light constitutes a sign of *expectancy* certified by every American myth of hospitality ("we'll leave the porch light on for you"). Therefore the left switch (*porch light*) up (*on*) means that company's coming (in the case at hand, *Martie for dinner and afterwards a movie*).

In the event, everyone decides to go. Getting ready to leave, they turn off all the lights downstairs (the family will not be receiving) except for the porch light (they all will be trooping out to the car). The last one out turns off the porch light. In the event, he forgets.

"Chanz? Would you go back and cut off the light?"

A second later, the house is dark.

RULELIST

Codes

tPRO   A rule that protects a thing
kPRO   A rule that protects a kid
tCON   A rule that is concerned with the control of a thing
kCON   A rule that is concerned with the control of a kid
tAPP   A rule that is concerned with the appearance of a thing (i.e., in relations, attitudes, orientations, style)
kAPP   A rule that is concerned with the appearance of a kid

Master List of All the Rules

1. "Don't push on the screen" (Chandler, tPRO).
2. "Don't push things through the holes in the screen" (Ingrid, tPRO).
3. "Don't slam it" (Chandler, ktAPP).
4. "Close it every time you go through" (Randall, tCON).
5. "Don't kick it" (Chandler, tPRO).
6. "Don't open it to strangers" (Randall, kCON).
7. "Don't open it to strangers or certain others" (Denis, kCON).
8. "Don't talk through the screen to guests or friends" (Denis, kAPP).
9. "Don't lock the screen against family members" (Denis, kCON).
10. "Don't slam the door" (Denis, ktAPP, tPRO).
11. "Don't play with the door" (Denis, kPRO, ktAPP, tPRO, tCON).
12. "Don't use the door as a 'gate' in a game; play inside or outside" (Ingrid, kPRO, tCON).
13. "Don't get your fingers caught in the door" (Denis, kPRO).
14. "Don't hang on the door" (Denis, tPRO).
15. "Don't kick it" (Chandler, tPRO).
16. "Don't open it too wide because it scratches the floor" (Chandler, tPRO).
17. "Don't let just anybody in" (Denis, kCON).

18. "Don't open the door to just anybody" (Denis, kCON).
19. "Don't go to the door and when you see somebody's there, just go away. Go up and open the door and see who it is" (Denis, kAPP).
20. "Close the door against the heat" (Ingrid, tCON).
21. "Close the door if the furnace is on" (Ingrid, tCON).
22. "Close the door behind you" (Denis, tCON).
23. "Open the door for fresh air" (Ingrid, tCON).
24. "Keep your hands off the woodwork" (Denis, ktAPP).
25. "Don't kick the doorframe" (Randall, kAPP).
26. "Keep your fingers *off* the windows" (Denis, ktAPP).
27. "Don't smudge up the windows" (Randall, ktAPP).
28. "Don't breathe on it" (Chandler, ktAPP).
29. "Don't do anything that will break the glass" (Ingrid, ktPRO).
30. "Don't breathe on them" (Chandler, ktAPP).
31. "Don't touch the windows" (Randall, ktAPP).
32. "Don't let your friends put their noses on the windows" (Denis, ktAPP).
33. "Don't wipe off the windows with your hands. You only smear them up" (Denis, ktAPP).
34. "Don't pound on them" (Chandler, kAPP, ktPRO).
35. "Don't spit on them" (Chandler, kCON).
36. "Don't play with the bells" (Denis, tCON).
37. "Don't dingdong them too much when someone's sitting in the living room" (Chandler, ktAPP).
38. "Keep your friends from dingdonging them too much" (Randall, ktAPP, tCON).
39. "Don't lock it" (Randall, ktAPP, kCON).
40. "Don't lock your friends out, or your brother" (Chandler, kCON).
41. "When you close the door, turn the handle . . . " (Denis, ktAPP, tCON, tPRO).
42. "Don't stamp on the floor" (Chandler, kAPP, kCON).
43. "Don't use it as a skating rink" (Randall, kAPP, ktPRO).
44. "Don't scratch up the floor" (Denis, tAPP, ktPRO).
45. "Don't mark the floor" (Denis, tAPP).
46. "Please don't spill on the floor" (Denis, tAPP).
47. "Pick up things you leave on the floor" (Ingrid, ktAPP, ktCON).
48. "Don't leave stuff on it so people can trip, like marbles" (Chandler, tCON, kPRO).
49. "There is no leaving of shoes around in this house" (Denis, kAPP, tCON).
50. "Make sure your feet are clean before you come in" (Denis, ktAPP).
51. "Don't walk on it with muddy feet" (Randall, ktAPP).
52. "You're not to climb on the radiator by yourself" (Denis, kPRO).

53. "We can't stand on it" (Randall, kPRO).
54. "Take items on the radiator upstairs when going up" (Ingrid, ktCON).
55. "Don't sit on it" (Chandler, ktPRO).
56. "Don't stand on it" (Randall, ktPRO).
57. "Don't bang into the newel post" (Denis, tPRO).
58. "Do you have to swing on the newel post like that?" (Denis, tPRO).
59. "Take up things on the newel post when you're going up" (Ingrid, ktCON).
60. "Don't leave anything on it" (Randall, tCON).
61. "Don't play around the banister" (Denis, kPRO).
62. "Don't slide down it" (Chandler, ktPRO).
63. "Don't play on the stairs" (Denis, ktCON, kPRO).
64. "Don't stomp on them" (Chandler, kAPP).
65. "Don't kick, mar, the risers" (Denis, ktAPP).
66. "Don't jump on the landing" (Denis, ktPRO).
67. "Don't stomp on it because the record will scratch" (Chandler, ktPRO).
68. "No playing on the landing" (Ingrid, tCON, kPRO).
69. "No leaving things on the landing" (Ingrid, tCON, kPRO).
70. "No playing on the stairs" (Denis, ktCON, kPRO).
71. "No running on the stairs, it makes too much noise" (Randall, kAPP, ktPRO).
72. "No clomping on the stairs" (Denis, kAPP, ktPRO).
73. "Stairs are for going up and down" (Ingrid, ktCON).
74. "Don't put your hands on the walls" (Chandler, tAPP).
75. "Don't mark up the walls" (Denis, ktAPP).
76. "Be careful of the walls" (Denis, tAPP).
77. "Don't kick them" (Chandler, ktAPP).
78. "Denis or Ingrid hang things on the walls" (Ingrid, tAPP, kCON, tPRO).
79. "Don't breathe on the windows" (Denis, ktAPP).
80. "Don't wipe off the windows with your hands" (Denis, ktAPP).
81. "Keep your hands/finger off the windows" (Denis, ktAPP).
82. "Don't smudge up the windows" (Denis, ktAPP).
83. "Don't mark up with hands" (Denis, ktAPP).
84. "Don't hurt the plants" (Denis, ktAPP, tPRO).
85. "Be careful of the plants" (Ingrid, tPRO).
86. "All plants, don't kill or injure—be nice" (Randall, kAPP, tPRO).
87. "Don't screw around with the easel" (Denis, kAPP, tCON, tPRO).
88. "Don't wind it up and down like we sometimes do" (Chandler, kAPP, tCON, tPRO).

89. "Denis puts the (chooses the) easel items except at Christmas when Ingrid decorates" (Ingrid, ktCON).
90. "Same as all pictures—don't put your hands on it (same as windows)" (Chandler, ktAPP, tPRO).
91. "Don't knock it off—never been told, but I assume that" (Randall, kAPP, tPRO).
92. "Don't touch it" (Denis, tPRO).
93. "Don't play with it" (Randall, kAPP, tPRO).
94. "Don't press the button when something is playing" (Chandler, kAPP, tPRO).
95. "Don't drop things on it from the stairs" (Chandler, kAPP, ktPRO).
96. "Don't play under it—like Transformers" (Chandler, kAPP, tPRO).
97. "Don't put stuff on top of the speaker" (Chandler, kAPP, tPRO).
98. "Cover speaker if you're doing work around it that might allow stuff to fall on it" (Ingrid, tPRO).
99. "Just like all the other plants—don't abuse it" (Randall, kAPP, tPRO).
100. "Don't abuse them—stereo used to be there" (Randall, tPRO).
101. "Don't leave anything on the cabinets" (Denis, ktAPP).
102. "Yes, that's a general rule, but the mail is placed there if you can't put it out" (Randall, kAPP, tCON).
103. "Be VERY careful" (Denis, tPRO).
104. "Don't play with it" (Chandler, tCON, tPRO).
105. "Don't put things in it" (Randall, tCON, tPRO).
106. "Be VERY careful" (Denis, tPRO).
107. "Don't spin unless you've made sure the wheel won't come off—and it's probably a good idea not to fool around with it at all" (Denis, tCON, tPRO).
108. "Don't play with it" (Randall, tPRO).
109. "Don't turn it, because it's broken" (Chandler, tPRO).
110. "Don't flip benches around" (Randall, kAPP, tPRO).
111. "Don't lay anything on them" (Randall, ktAPP).
112. "Put records back where you got them" (Denis, tCON).
113. "Put them away after use—" (Chandler, kAPP, tCON).
114. "Unless you don't know where they go" (Randall, tCON).
115. "Don't leave them around" (Denis, kAPP, tPRO).
116. "Make sure you put sleeves on records so that opening is up" (Randall, tCON, tPRO).
117. "Be careful not to scratch" (Randall, tPRO).
118. "Be careful in general" (Chandler, ktCON, tPRO).
119. "Don't use them as Frisbees" (Randall, kAPP).
120. "Put them back when you're done" (Randall, ktCON).

121. "Don't leave them out when you're done" (Chandler, kAPP).
122. "Don't pull them out by the top of the spine—" (Denis, tPRO) "—right, by the middle of the spine" (Randall, tPRO).
123. "Treat with care" (Denis, ktCON, tPRO).
124. "Treat it like a chair, not like a toy" (Denis, kAPP, ktCON).
125. "Don't bounce on it" (Chandler, kAPP, ktCON, tPRO).
126. "You fall more easily than on a normal chair if you stand on it" (Randall, kPRO).
127. "Put it back when you're done" (Chandler, ktCON).
128. "If you bicker over who'll sit on it you can't use it" (Ingrid, kAPP).
129. "Don't throw yourself onto the couch—sit down on it" (Denis, kAPP, ktCON, tPRO).
130. "Don't sit on the arms" (Chandler, Randall, kAPP, tPRO).
131. "Don't sit on the back" (Chandler, kAPP, tPRO).
132. "Don't hang off back of couch" (Denis, ktAPP, tPRO).
133. "Don't mark back with shoes" (Chandler, ktAPP, tPRO).
134. "Don't climb over the couch" (Chandler, ktAPP, tPRO).
135. "No fighting on the couch" (Denis, ktAPP).
136. "No shoes on the couch" (Chandler, ktAPP, tPRO).
137. "Don't put food on it" (Chandler, ktAPP, tPRO).
138. "Don't throw up on it—go outdoors or in the bathroom" (Randall, tAPP, kCON).
139. "Don't get the couch dirty" (Denis, tAPP).
140. "Don't hide things under or behind the pillows" (Randall, kCON).
141. "Don't take up the pillows and sit underneath them" (Chandler, Randall, kAPP, tCON).
142. "Don't leave the pillows not spiffed up" (Randall, tAPP, kCON).
143. "Don't leave things under the couch" (Chandler, ktAPP).
144. "Don't leave your shoes under the couch" (Denis, kAPP, tCON).
145. "Don't touch—just look at it" (Chandler, kCON, tPRO).
146. "Don't play with it" (Chandler, kAPP, tPRO).
147. "Keep fingers off" (Denis, tAPP).
148. "Don't breathe on bottom" (Randall, tAPP).
149. "Don't tilt the shade" (Randall, ktAPP).
150. "Don't take the shade off and use it as a space helmet" (Randall, kAPP).
151. "The lamp can be moved, but you have to put it back where it goes" (Denis, ktCON).
152. "Don't leave it on all night" (Randall, Chandler, tCON).
153. "Don't play around with this" (Randall, tPRO).
154. "Don't put your feet on this chair" (Chandler, ktAPP).

155. "No standing on the rockers" (Ingrid, ktPRO).
156. "Don't play around with the wicker" (Randall, kAPP, tPRO).
157. "Don't kick it" (Randall, kAPP, tPRO).
158. "Be real careful with this one because we destroyed the last one" (Randall, kCON, tPRO).
159. "No throwing the throw pillows" (Denis, kAPP, ktCON, tPRO).
160. "No whacking people with them" (Randall, kAPP, kCON).
161. "No pillow fights" (Denis, ktCON).
162. "Don't leave them on the floor" (Chandler, kAPP, tCON).
163. "Straighten out pillows after use" (Denis, tAPP, kCON).
164. "Be careful" (Denis, tCON).
165. "Please don't move them" (Denis, tCON).
166. "But we move them at Christmas" (Chandler, tCON, tPRO).
167. "Don't unplug wires or fiddle with knob on back" (Chandler, tCON, tPRO).
168. "Don't take off covers" (Chandler, tPRO).
169. "Don't touch the speakers under the cover if you take them off" (Chandler, tPRO).
170. "Don't leave things on it" (Chandler, ktAPP).
171. "Don't play the records too loud: if you do, make sure Bill's window is shut" (Denis, ktAPP, ktCON, tPRO).
172. "Unless Den's working, then we can't play it too loud" (Chandler, ktCON).
173. "Under no circumstances touch this drum" (Denis, tPRO).
174. "Don't play with it" (Randall, tPRO).
175. "Don't put marbles or anything in the cup part" (Chandler, tCON, tPRO).
176. "Don't play with it—you could knock it over" (Chandler, kAPP, tPRO).
177. "Don't touch it unless to turn it on or off" (Randall, ktCON).
178. "Don't take off the lamp shade and use it as a space helmet" (Chandler, kAPP, tPRO).
179. "Don't move it" (Randall, tCON, tPRO).
180. "Don't breathe on base" (Randall, tAPP).
181. "Don't hurt it" (Denis, kAPP, tPRO).
182. "Don't abuse it" (Randall, kAPP, tPRO).
183. "Don't be mean to it" (Chandler, kAPP, tPRO).
184. "Don't step on it" (Chandler, Randall, kAPP, tPRO).
185. "I think—don't take it out and put it someplace else" (Chandler, tCON).
186. "Don't stand on the radiator" (Randall, ktPRO).
187. "Don't put stuff on it and leave it there" (Randall, kAPP, ktCON).
188. "Keep your fingers off the glass" (Denis, ktAPP).

189. "Make sure glass and screen don't get broken" (Ingrid, ktPRO).
190. "Don't open like *boom-boom:* open and close continuously" (Randall, ktCON).
191. "Don't try to talk through the window" (Denis, kAPP, kCON).
192. "Don't let friends press their heads against it" (Randall, kAPP, kCON, tPRO).
193. "Close the storms if you're washing the porch" (Ingrid, kCON).
194. "When playing a record, close the window" (Chandler, ktAPP, ktCON).
195. "Close the window when it starts to rain because the front windows are protected by the porch" (Chandler, ktCON, tPRO).
196. "Don't sit down on it real hard" (Randall, kAPP, ktCON, tPRO).
197. "Don't bounce on it" (Randall, kAPP, ktCON, tPRO).
198. "Don't hang on back" (Chandler, kAPP, tPRO).
199. "Don't stand on it" (Randall, ktPRO).
200. "Don't untie the leather" (Chandler, ktCON, tPRO).
201. "Not supposed to play or mess around with the king" (Randall, tPRO).
202. "We shouldn't play in it" (Chandler, kAPP).
203. "Don't pull the chain" (Chandler, ktAPP, tCON).
204. "We shouldn't wind it up and play with it" (Chandler, Randall, tPRO).
205. "Don't play around with it" (Chandler, Randall, ktPRO).
206. "Don't stick fingers in it or anything like that" (Chandler, Randall, kPRO).
207. "Don't cut the fan off and on, off and on" (Denis, kAPP, kCON).
208. "Don't touch the woodwork with your dirty finger" (Denis, tAPP).
209. "Don't hang off the mantle" (Denis, tPRO).
210. "Don't play around with anything on the mantle because you could break some of them" (Randall, kCON, tPRO).
211. "Try to keep your fingers off the table" (Denis, ktAPP).
212. "Don't play with anything that will scratch it up" (Chandler, ktPRO).
213. "Don't put your feet on the table" (Chandler, kAPP, kCON).
214. "Don't leave your stuff on the table" (Denis, ktAPP).
215. "Don't leave *Tintins* on the table" (Randall, kAPP).
216. "Don't lift glass part off" (Randall, tPRO).
217. "We probably shouldn't play with it, but we do" (Randall, ktCON).
218. "Don't touch it—we'll get it dirty" (Randall, ktAPP).
219. "Don't put hand up on it" (Randall, ktAPP).
220. "Don't throw anything at it" (Denis, ktAPP).
221. "Leave the light switches the way they are" (Denis, tCON).
222. "Don't play with the light switches" (Denis, kAPP, ktCON).
223. "Don't mess with one of the switches, because it turns off the upstairs light" (Randall, kAPP, tCON).

Master List of PROtection Rules

1. "Don't push on the screen" (Chandler, tPRO).
2. "Don't push things through the holes in the screen" (Ingrid, tPRO).
5. "Don't kick it" (Chandler, tPRO).
10. "Don't slam the door" (Denis, ktAPP, tPRO).
11. "Don't play with the door" (Denis, kPRO, ktAPP, tPRO, tCON).
12. "Don't use the door as a 'gate' in a game: play inside or outside" (Ingrid, kPRO, tCON).
13. "Don't get your fingers caught in the door" (Denis, kPRO).
14. "Don't hang on the door" (Denis, tPRO).
15. "Don't kick it" (Chandler, tPRO).
16. "Don't open it too wide because it scratches the floor" (Chandler, tPRO).
29. "Don't do anything that will break the glass" (Ingrid, ktPRO).
34. "Don't pound on them" (Chandler, kAPP, ktPRO).
41. "When you close the door, turn the handle . . ." (Denis, ktAPP, tPRO).
43. "Don't use it as a skating rink" (Randall, kAPP, ktPRO).
44. "Don't scratch up the floor" (Denis, tAPP, ktPRO).
48. "Don't leave stuff on it so people can trip, like marbles" (Chandler, tCON, kPRO).
52. "You're not to climb on the radiator by yourself" (Denis, kPRO).
53. "We can't stand on it" (Randall, kPRO).
55. "Don't sit on it" (Chandler, ktPRO).
56. "Don't stand on it" (Randall, ktPRO).
57. "Don't bang into the newel post" (Denis, tPRO).
58. "Do you have to swing on the newel post like that?" (Denis, tPRO).
61. "Don't play around the banister" (Denis, kPRO).
62. "Don't slide down it" (Chandler, ktPRO).
63. "Don't play on the stairs" (Denis, ktCON, kPRO).
66. "Don't jump on the landing" (Denis, ktPRO).
67. "Don't stomp on it because the record will scratch" (Chandler, ktPRO).
68. "No playing on the landing" (Ingrid, tCON, kPRO).
69. "No leaving things on the landing" (Ingrid, tCON, kPRO).
70. "No playing on the stairs" (Denis, ktCON, kPRO).
71. "No running on the stairs, it makes too much noise" (Randall, kAPP, ktPRO).
72. "No clomping on the stairs" (Denis, kAPP, ktPRO).
78. "Denis or Ingrid hang things on the walls" (Ingrid, tAPP, kCON, tPRO).
84. "Don't hurt the plants" (Denis, ktAPP, tPRO).
85. "Be careful of the plants" (Ingrid, tPRO).

86. "All plants, don't kill or injure—be nice" (Randall, kAPP, tPRO).
87. "Don't screw around with the easel" (Denis, kAPP, tCON, tPRO).
88. "Don't wind it up and down like we sometimes do" (Chandler, kAPP, tCON, tPRO).
90. "Same as all pictures—don't put your hands on it (same as windows)" (Chandler, ktAPP, tPRO).
91. "Don't knock it off—never been told, but I assume that" (Randall, kAPP, tPRO).
92. "Don't touch it" (Denis, tPRO).
93. "Don't play with it" (Randall, kAPP, tPRO).
94. "Don't press the button when something is playing" (Chandler, kAPP, tPRO).
95. "Don't drop things on it from the stairs" (Chandler, kAPP, ktPRO).
96. "Don't play under it—like Transformers" (Chandler, kAPP, tPRO).
97. "Don't put stuff on top of the speaker" (Chandler, kAPP, tPRO).
98. "Cover speaker if you're doing work around it that might allow stuff to fall on it" (Ingrid, tPRO).
99. "Just like all the other plants—don't abuse it" (Randall, kAPP, tPRO).
100. "Don't abuse them—stereo used to be there" (Randall, tPRO).
103. "Be VERY careful" (Denis, tPRO).
104. "Don't play with it" (Chandler, tCON, tPRO).
105. "Don't put things in it" (Randall, tCON, tPRO).
106. "Be VERY careful" (Denis, tPRO).
107. "Don't spin unless you're made sure the wheel won't come off—and it's probably a good idea not to fool around with it at all" (Denis, tCON, tPRO).
108. "Don't turn it, because it's broken" (Chandler, tPRO).
109. "Don't play with it" (Randall, tPRO).
110. "Don't flip benches around" (Randall, kAPP, tPRO).
115. "Don't leave them around" (Denis, kAPP, tPRO).
116. "Make sure you put sleeves on records so that opening is up" (Randall, tCON, tPRO).
117. "Be careful not to scratch" (Randall, tPRO).
118. "Be careful in general" (Chandler, ktCON, tPRO).
122. "Don't pull them out by the top of the spine—" (Denis, tPRO) "right, by the middle of spine" (Randall, tPRO).
123. "Treat with care" (Denis, ktCON, TPRO).
126. "You fall more easily than on a normal chair if you stand on it" (Randall, kPRO).
130. "Don't sit on the arms" (Chandler, Randall, kAPP, tPRO).
131. "Don't sit on the back" (Chandler, kAPP, TPRO).
132. "Don't hang off back of couch" (Denis, ktAPP, tPRO).

133. "Don't mark back with shoes" (Chandler, ktAPP, tPRO).

134. "Don't climb over the couch" (Chandler, kAPP, tPRO).

136. "No shoes on the couch" (Chandler, ktAPP, tPRO).

137. "Don't put food on it" (Chandler, ktAPP, tPRO).

146. "Don't play with it" (Chandler, kAPP, tPRO).

153. "Don't play around with this" (Randall, tPRO).

155. "No standing on the rockers" (Ingrid, ktPRO).

156. "Don't play around with the wicker" (Randall, kAPP, tPRO).

157. "Don't kick it" (Randall, kAPP, tPRO).

158. "Be real careful with this one, because we destroyed the last one" (Randall, kCON, tPRO).

159. "No throwing the throw pillows" (Denis, kAPP, ktCON, tPRO).

164. "Be careful" (Denis, tPRO).

166. "But we move them at Christmas" (Chandler, tCON, tPRO).

167. "Don't unplug wires or fiddle with knob on back" (Chandler, tCON, tPRO).

168. "Don't take off covers" (Chandler, tPRO).

169. "Don't touch the speakers under the cover if you take them off" (Chandler, tPRO).

171. "Don't play the records too loud: if you do make sure Bill's window is shut" (Denis, ktAPP, ktCON, tPRO).

173. "Under no circumstances touch this drum" (Denis, tPRO).

174. "Don't play with it" (Randall, tPRO).

175. "Don't put marbles or anything in the cup part" (Chandler, tCON, tPRO).

176. "Don't play with it—you could knock it over" (Chandler, kAPP, tPRO).

178. "Don't take off the lamp shade and use it as a space helmet" (Chandler, kAPP, tPRO).

179. "Don't move it" (Randall, tCON, tPRO).

181. "Don't hurt it" (Denis, kAPP, tPRO).

182. "Don't abuse it" (Randall, kAPP, tPRO).

183. "Don't be mean to it" (Chandler, kAPP, tPRO).

184. "Don't step on it" (Chandler, Randall, kAPP, tPRO).

186. "Don't stand on the radiator" (Randall, ktPRO).

189. "Make sure glass and screen don't get broken" (Ingrid, ktPRO).

192. "Don't let friends press their heads against it" (Randall, kAPP, kCON, tPRO).

195. "Close the window when it starts to rain because the front windows are protected by the porch" (Chandler, ktCON, tPRO).

196. "Don't sit down on it real hard" (Randall, kAPP, ktCON, tPRO).

197. "Don't bounce on it" (Randall, kAPP, ktCON, tPRO).

198. "Don't hang on back" (Chandler, kAPP, tPRO).

199. "Don't stand on it" (Randall, ktPRO).
200. "Don't untie the leather" (Chandler, ktCON, tPRO).
201. "Not supposed to play or mess around with the king" (Randall, tPRO).
204. "We shouldn't wind it up and play with it" (Chandler, Randall, tPRO).
205. "Don't play around with it" (Chandler, Randall, ktPRO).
206. "Don't stick fingers in it or anything like that" (Chandler, Randall, kPRO).
209. "Don't hang off the mantle" (Denis, tPRO).
210. "Don't play around with anything on the mantle because you could break some of them" (Randall, kCON, tPRO).
212. "Don't play with anything that will scratch it up" (Chandler, ktPRO).
216. "Don't lift glass part off" (Randall, tPRO).

Master List of CONtrol Rules

4. "Close it every time you go through" (Randall, tCON).
6. "Don't open it to strangers" (Randall, kCON).
7. "Don't open it to strangers or certain others" (Denis, kCON).
9. "Don't lock the screen against family members" (Denis, kCON).
11. "Don't play with the door" (Denis, kPRO, ktAPP, tPRO, tCON).
12. "Don't use the door as a 'gate' in a game: play inside or outside" (Ingrid, kPRO, tCON).
17. "Don't let just anybody in" (Denis, kCON).
18. "Don't open the door to just anybody" (Denis, kCON).
20. "Close the door against the heat" (Ingrid, tCON).
21. "Close the door if the furnace is on" (Ingrid, tCON).
22. "Close the door behind you" (Denis, tCON).
23. "Open the door for fresh air" (Ingrid, tCON).
35. "Don't spit on them" (Chandler, kCON).
36. "Don't play with the bells" (Denis, tCON).
38. "Keep your friends from dingdonging them too much" (Randall, ktAPP, tCON).
39. "Don't lock it" (Randall, ktAPP, kCON).
40. "Don't lock your friends out, or your brother" (Chandler, kCON).
41. "When you close the door, turn the handle . . ." (Denis, ktAPP, tCON, tPRO).
42. "Don't stamp on the floor" (Chandler, kAPP, kCON).
47. "Pick up things you leave on the floor" (Ingrid, ktAPP, ktCON).
48. "Don't leave stuff on it so people can trip, like marbles" (Chandler, tCON, kPRO).
49. "There is no leaving of shoes around in this house" (Denis, kAPP, tCON).

54. "Take items on the radiator upstairs when going up" (Ingrid, ktCON).

59. "Take up things on the newel post when you're going up" (Ingrid, ktCON).

60. "Don't leave anything on it" (Randall, tCON).

63. "Don't play on the stairs" (Denis, ktCON, kPRO).

68. "No playing on the landing" (Ingrid, tCON, kPRO).

69. "No leaving things on the landing" (Ingrid, tCON, kPRO).

70. "No playing on the stairs" (Denis, ktCON, kPRO).

73. "Stairs are for going up and down" (Ingrid, ktCON).

78. "Denis or Ingrid hang things on the walls" (Ingrid, tAPP, kCON, tPRO).

87. "Don't screw around with the easel" (Denis, kAPP, tCON, tPRO).

88. "Don't wind it up and down like we sometimes do" (Chandler, kAPP, tCON, tPRO).

89. "Denis puts the (chooses the) easel items except at Christmas when Ingrid decorates" (Ingrid, ktCON).

102. "Yes, that's a general rule, but the mail is placed there if you can't put it out" (Randall, kAPP, tCON).

104. "Don't play with it" (Chandler, tCON, tPRO).

105. "Don't put things in it" (Randall, tCON, tpro).

107. "Don't spin unless you're made sure the wheel won't come off—and it's probably a good idea not to fool around with it at all" (Denis, tCON, tPRO).

112. "Put records back where you got them" (Denis, tCON).

113. "Put them away after use—" (Chandler, kAPP, tCON).

114. "Unless you don't know where they go" (Randall, tCON).

116. "Make sure you put sleeves on records so that opening is up" (Randall, tCON, tPRO).

118. "Be careful in general" (Chandler, ktCON, tPRO).

120. "Put them back when you're done" (Randall, ktCON).

123. "Treat them with care" (Denis, ktCON, tPRO).

124. "Treat it like a chair, not like a toy" (Denis, kAPP, ktCON).

125. "Don't bounce on it" (Chandler, kAPP, ktCON, tPRO).

127. "Put it back when you're done" (Chandler, ktCON).

129. "Don't throw yourself onto the couch—sit down on it" (Denis, kAPP, ktCON, tPRO).

138. "Don't throw up on it—go outdoors or in the bathroom" (Randall, tAPP, kCON).

140. "Don't hide things under or behind the pillows" (Randall, kCON).

141. "Don't take up the pillows and sit underneath them" (Chandler, Randall, kAPP, tCON).

142. "Don't leave the pillows not spiffed up" (Randall, tAPP, kCON).

144. "Don't leave your shoes under the couch" (Denis, kAPP, tCON).
145. "Don't touch—just look at it" (Chandler, kCON, tPRO).
151. The lamp can be moved "but you have to put it back where it goes" (Denis, ktCON).
152. "Don't leave it on all night" (Randall, Chandler, tCON).
158. "Be careful with this one, because we destroyed the last one" (Randall, kCON, tPRO).
159. "No throwing the throw pillows" (Denis, kAPP, ktCON, tPRO).
160. "No whacking people with them" Randall, kAPP, kCON).
161. "No pillow fights" (Denis, ktCON).
162. "Don't leave them on the floor" (Chandler, kAPP, tCON).
163. "Straighten out pillows after use" (Denis, tAPP, kCON).
165. "Please don't move them" (Denis, tCON).
166. "But we move them at Christmas" (Chandler, tCON, tPRO).
167. "Don't unplug wires or fiddle with knob on back" (Chandler, tCON, tPRO).
171. "Don't play the records too loud: if you do make sure Bill's window is shut" (Denis, ktAPP, ktCON, tPRO).
172. "Unless Den's working, then we can't play it too loud" (Chandler, ktCON).
175. "Don't put marbles or anything in the cup part" (Chandler, tCON, tPRO).
177. "Don't touch it unless to turn it on or off" (Randall, ktCON).
179. "Don't move it" (Randall, tCON, tPRO).
185. "I think—don't take it out and put it someplace else" (Chandler, tCON).
187. "Don't put stuff on it and leave it there" (Randall, kAPP, ktCON).
190. "Don't open like *boom-boom:* open and close continuously" (Randall, ktCON).
191. "Don't try to talk through the window" (Denis, kAPP, kCON).
192. "Don't let friends press their heads against it" (Randall, kAPP, kCON, tPRO).
193. "Close the storms if you're washing the porch" (Ingrid, kCON).
194. "When playing a record, close the window" (Chandler, ktAPP, ktCON).
195. "Close the window when it starts to rain because the front windows are protected by the porch" (Chandler, ktCON, tPRO).
196. "Don't sit down on it real hard" (Randall, kAPP, ktCON, tPRO).
197. "Don't bounce on it" (Randall, kAPP, ktCON, tPRO).
200. "Don't untie the leather" (Chandler, ktCON, tPRO).
203. "Don't pull the chain" (Chandler, ktAPP, tCON).
207. "Don't cut the fan off and on, off and on" (Denis, kAPP, kCON).
210. "Don't play around with anything on the mantle because you could break some of them" (Randall, kCON, tPRO).
213. "Don't put your feet on the table" (Chandler, kAPP, kCON).
217. "We probably shouldn't play with it, but we do" (Randall, ktCON).

221. "Leave the light switches the way they are" (Denis, tCON).
222. "Don't play with the light switches" (Denis, kAPP, ktCON).
223. "Don't mess with one of the switches, because it turns off the upstairs light" (Randall, kAPP, tCON).

Master List of APPearance Rules

3. "Don't slam it" (Chandler, ktAPP).
8. "Don't talk through the screen to guests or friends" (Denis, kAPP).
10. "Don't slam the door" (Denis, ktAPP, tPRO).
11. "Don't play with the door" (Denis, kPRO, ktAPP, tPRO, tCON).
19. "Don't go to the door and when you see somebody's there, just go away. Go up and open the door and see who it is" (Denis, kAPP).
24. "Keep your hands off the woodwork" (Denis, ktAPP).
25. "Don't kick the doorframe" (Randall, kAPP).
26. "Keep your fingers *off* the windows" (Denis, ktAPP).
27. "Don't smudge up the windows" (Randall, ktAPP).
28. "Don't breathe on it" (Chandler, ktAPP).
30. "Don't breathe on them" (Chandler, ktAPP).
31. "Don't touch the windows" (Randall, ktAPP).
32. "Don't let your friends put their noses on the windows" (Denis, ktAPP).
33. "Don't wipe off the windows with your hands. You only smear them up" (Denis, ktAPP).
34. "Don't pound on them" (Chandler, kAPP, ktPRO).
37. "Don't dingdong them too much when someone's sitting in the living room" (Chandler, ktAPP).
38. "Keep your friends from dingdonging them too much" (Randall, ktAPP, tCON).
39. "Don't lock it" (Randall, ktAPP, kCON).
41. "When you close the door, turn the handle . . ." (Denis, ktAPP, tPRO).
42. "Don't stamp on the floor" (Chandler, kAPP, kCON).
43. "Don't use it as a skating rink" (Randall, kAPP, ktPRO).
44. "Don't scratch up the floor" (Denis, tAPP, ktPRO).
45. "Don't mark the floor" (Denis, tAPP).
46. "Please don't spill on the floor" (Denis, tAPP).
47. "Pick up things you leave on the floor" (Ingrid, ktAPP, ktCON).
49. "There is no leaving of shoes around in this house" (Denis, kAPP, tCON).
50. "Make sure your feet are clean before you come in" (Denis, ktAPP).

51. "Don't walk on it with muddy feet" (Randall, ktАРР).
64. "Don't stomp on them" (Chandler, kАРР).
65. "Don't kick, mar, the risers" (Denis, tАРР).
71. "No running on the stairs, it makes too much noise" (Randall, kАРР, ktРRO).
72. "No clomping on the stairs" (Denis, kАРР, ktРRO).
74. "Don't put your hands on the walls" (Chandler, tАРР).
75. "Don't mark up the walls" (Denis, ktАРР).
76. "Be careful of the walls" (Denis, tАРР).
77. "Don't kick them" (Chandler, ktАРР).
78. "Denis or Ingrid hang things on the walls" (Ingrid, tАРР, kСON, tРRO).
79. "Don't breathe on the windows" (Denis, ktАРР).
80. "Don't wipe off the windows with your hands" (Denis, ktАРР).
81. "Keep your hands/finger off the windows" (Denis, ktАРР).
82. "Don't smudge up the windows" (Randall, ktАРР).
83. "Don't mark up with hands" (Chandler, ktАРР).
84. "Don't hurt the plants" (Denis, ktАРР, tРRO).
86. "All plants, don't kill or injure—be nice" (Randall, kАРР, tРRO).
87. "Don't screw around with the easel" (Denis, kАРР, tСON, tРRO).
88. "Don't wind it up and down like we sometimes do" (Chandler, kАРР, tСON, tРRO).
90. "Same as all pictures—don't put your hands on it (same as windows)" (Chandler, ktАРР, tРRO).
91. "Don't knock it off—never been told, but I assume that" (Randall, kАРР, tРRO).
93. "Don't play with it" (Randall, kАРР, tРRO).
94. "Don't press the button when something is playing" (Chandler, kАРР, tРRO).
95. "Don't drop things on it from the stairs" (Chandler, kАРР, ktРRO).
96. "Don't play under it—like Transformers" (Chandler, kАРР, tРRO).
97. "Don't put stuff on top of the speaker" (Chandler, kАРР, tРRO).
99. "Just like all the other plants—don't abuse it" (Randall, kАРР, tРRO).
101. "Don't leave anything on the cabinets" (Denis, ktАРР).
102. "Yes, that's a general rule, but the mail is placed there if you can't put it out" (Randall, kАРР, tСON).
110. "Don't flip benches around" (Randall, kАРР, tРRO).
111. "Don't lay anything on them" (Randall, ktАРР).
115. "Don't leave them around" (Denis, kАРР, tРRO).
119. "Don't use them as Frisbees" (Randall, kАРР).
121. "Don't leave them out when you're done" (Chandler, kАРР).
124. "Treat it like a chair, not like a toy" (Denis, kАРР, ktСON).

125. "Don't bounce on it" (Chandler, kAPP, ktCON, tPRO).
128. "If you bicker over who'll sit on it you can't use it" (Ingrid, kAPP).
129. "Don't throw yourself onto the couch—sit down on it" (Denis, kAPP, ktCON, tPRO).
130. "Don't sit on the arms" (Chandler, Randall, kAPP, tPRO).
131. "Don't sit on the back" (Chandler, kAPP, tPRO).
132. "Don't hang off back of couch" (Denis, ktAPP, tPRO).
133. "Don't mark back with shoes" (Chandler, ktAPP, tPRO).
134. "Don't climb over the couch" (Chandler, kAPP, tPRO).
135. "No fighting on the couch" (Denis, ktAPP).
136. "No shoes on the couch" (Chandler, ktAPP, tPRO).
137. "Don't put food on it" (Chandler, ktAPP, tPRO).
138. "Don't throw up on it—go outdoors or in the bathroom" (Randall, tAPP, kCON).
139. "Don't get the couch dirty" (Denis, tAPP).
141. "Don't take up the pillows and sit underneath them" (Chandler, Randall, kAPP, tCON).
142. "Don't leave the pillows not spiffed up" (Randall, tAPP, kCON).
143. "Don't leave things under the couch" (Chandler, ktAPP).
144. "Don't leave your shoes under the couch" (Denis, kAPP, tCON).
146. "Don't play with it" (Chandler, kAPP, tPRO).
147. "Keep fingers off" (Denis, tAPP).
148. "Don't breathe on bottom" (Randall, tAPP).
149. "Don't tilt the shade" (Randall, ktAPP).
150. "Don't take the shade off and use it as a space helmet" (Randall, kAPP).
154. "Don't put your feet on this chair" (Chandler, ktAPP).
156. "Don't play around with the wicker" (Randall, kAPP, tPRO).
157. "Don't kick it" (Randall, kAPP, tPRO).
159. "No throwing the throw pillows" (Denis, kAPP, ktCON, tPRO).
160. "No whacking people with them" (Randall, kAPP, kCON).
162. "Don't leave them on the floor" (Chandler, kAPP, tCON).
163. "Straighten out pillows after use" (Denis, tAPP, kCON).
170. "Don't leave things on it" (Chandler, ktAPP).
171. "Don't play the records too loud: if you do make sure Bill's window is shut" (Denis, ktAPP, ktCON, tPRO).
176. "Don't play with it—you could knock it over" (Chandler, kAPP, tPRO).
178. "Don't take off the lamp shade and use it as a space helmet" (Chandler, kAPP, tPRO).
180. "Don't breathe on base" (Randall, tAPP).

181. "Don't hurt it" (Denis, kAPP, tPRO).
183. "Don't be mean to it" (Chandler, kAPP, tPRO).
184. "Don't step on it" (Chandler, Randall, kAPP, tPRO).
187. "Don't put stuff on it and leave it there" (Randall, kAPP, ktCON).
188. "Keep your fingers off the glass" (Denis, ktAPP).
191. "Don't try to talk through the window" (Denis, kAPP, kCON).
192. "Don't let friends press their heads against it" (Randall, kAPP, kCON, tPRO).
194. "When playing a record, close the window" (Chandler, ktAPP, ktCON).
196. "Don't sit down on it real hard" (Randall, kAPP, ktCON, tPRO).
197. "Don't bounce on it" (Randall, kAPP, ktCON, tPRO).
198. "Don't hang on back" (Chandler, kAPP, tPRO).
202. "We shouldn't play in it" (Chandler, kAPP).
203. "Don't pull the chain" (Chandler, ktAPP, tCON).
207. "Don't cut the fan off and on, off and on" (Denis, kAPP, kCON).
208. "Don't touch the woodwork with your dirty finger" (Denis, tAPP).
211. "Try to keep your fingers off the table" (Denis, ktAPP).
213. "Don't put your feet on the table" (Chandler, kAPP, kCON).
214. "Don't leave your stuff on the table" (Denis, ktAPP).
215. "Don't leave *Tintins* on the table" (Randall, kAPP).
218. "Don't touch it—we'll get it dirty" (Randall, ktAPP).
219. "Don't put hand up on it" (Randall, ktAPP).
220. "Don't throw anything at it" (Denis, ktAPP).
222. "Don't play with the light switches" (Denis, kAPP, ktCON).
223. "Don't mess with one of the switches, because it turns off the upstairs light" (Randall, kAPP, tCON).

Hand-offs

1. "Same as small ones" (Randall, to RULES 100, 101, 102)
2. "Same as every painting" (Chandler, Randall, to RULES 90, 91, 145).
3. "Same as others" (Chandler, Randall, to RULES 52, 53, 186, 187).
4. "Same as other windows" (Chandler, Randall, to RULES 26, 27, 28, 29, 30, 31, 32, 33, 34, 35, 79, 80, 81, 82, 83, 188, 189, 190, 191).
5. "Same as the others" (Chandler, Randall, to RULES 1, 2, 192).
6. "Same as other speakers" (Chandler, to RULES 164, 165, 167, 168, 169, 170, 171, 172).
7. "Just like all the plants" (Chandler, Randall, to RULES 84, 85, 86, 99, 181, 182, 183, 184).

Unspoken Rules

A. "Don't throw things in the living room—you might hit something" (Ingrid, kAPP, kCON, tPRO).
B. "Walk in the house!" (Ingrid, kCON).
C. "Keep the house safe" (Ingrid, ktPRO).
D. "Explain to your friends what the rules of the house are" (Ingrid, kCON).
E. "Don't treat the room as a jungle gym" (Randall, kAPP, kCON, ktPRO).

INDEX OF PARADIGMS

head/stomach, 107
hidden/revealed, 56, 61
high/low, 43
high end/low end, 51
high-fired/low-fired, 294
high lead oxide content/low lead oxide content, 46
high order-energy/low order-energy, 294
high social status/low social status, 46
holidays/work days, 147
housing projects/suburban apartment, 159
humble/proud, 134

Ingrid's side of the family/Denis's side of the family, 46, 48
inner city/suburb, 115

kitchen table/dining room table/living room table, 209

large/small, 46
left/middle/right, 304
left/right, 294
leisure/work, 147
like an architectural magazine/not like an architectural magazine, 209
living room/dining room–kitchen, 81
living room/kitchen, 147, 247
living room/rest of house, 48
living room–dining room/kitchen, 81
living room–dining room/rest of house, 48
loud/quiet, 43

machine/man/nature, 43, 229
machine/———/man/nature, 43
male/female, 247
material ambition/dreamy-ecstatic, 254
matte/shiny, 60
measurable accomplishment/waste of time, 254–55

medieval/renaissance/baroque/classical/romantic/modern, 148
metal/wood, 291
mind/hand, 147
Motown/California sound/British invasion, 148
mouth/anus, 48, 50, 115
mud/grime/dirt/soot/dust, 227

Nature/Culture, 60, 229, 294
noble/common, 292, 294
noble-rare/common, 294
noise/quiet, 25
Norman Rockwell/surrealists–abstract expressionists, 260
nuclear family/tribe, 254

of regional origin/of national origin, 294
of the house/of the Family Wood, 46
old/new, 46, 289, 294
old fashioned/modern, 43, 46
on/off, 308
ordered/confused, 43
orthogonal grid/curved streets, 115

painting/drawing/print/photograph, 292
painting/photograph, 292, 294
perverse obscenity/Mrs. Grundy, 207
place of residence/place of work, 147
plant/mineral, 134
primitive/folk, 134, 294
"primitive"/peasant/industrial, 208
primitive/sophisticated, 203, 294
propertied/not propertied, 294
public/private, 75, 77

rare/common, 292, 294
rational thought/pure psychic automatism, 204, 206
raw/cooked, 60, 304
realistic/imaginative, 292

Denis Wood was born and raised in Cleveland, Ohio, where he graduated from Western Reserve University with a degree in English. He received his master's and doctorate degrees in geography from Clark University in Worcester, Massachusetts, where he subsequently taught high school. Since 1973 he has taught environmental psychology and design at the School of Design at North Carolina State University, where he is Professor of Design. Author of *The Power of Maps,* he also cocurated the Cooper-Hewitt National Museum of Design exhibition of the same name (subsequently mounted at the Smithsonian in Washington). He and Robert Beck have worked together on issues of human development for twenty years. They have previously published joint work in psychogeography, environmental psychology, children's environments, and dialogue processes.

Robert J. Beck, a native of Brooklyn, N.Y., was raised in Southern California and received his Ph.D. from the Committee on Human Development at the University of Chicago in 1965. Following postdoctoral training in psychoanalysis, Beck was in private practice in New York City and Montreal for 20 years. As an environmental psychologist at Clark University and l'Université de Montréal, he published research on mental maps, housing, and hospital design. Since 1987 Beck has taught educational psychology at the City College of New York and conducted research on children's moral and narrative development. Producer/director of the award-winning *Video Dictionary of Classical Ballet,* Beck is currently Research Fellow in arts and education, Meadows School of the Arts, Southern Methodist University.

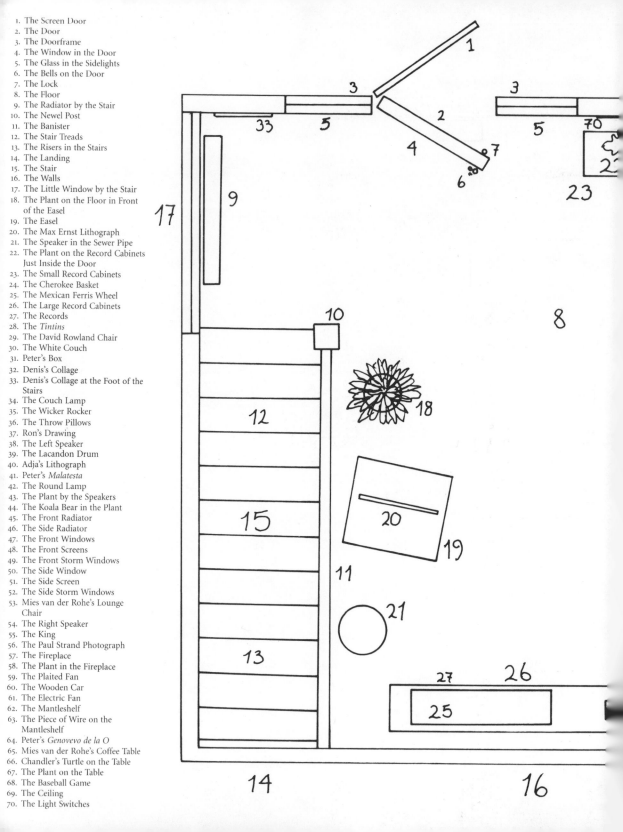